DICTIONARY OF SPACE
and Astronomy

Published by Granada Publishing 1985
Granada Publishing Limited
8 Grafton Street, London W1X 3LA

Copyright © Granada Publishing 1985

British Library Cataloguing in Publication Data

Kerrod, Robin
  Dictionary of space and astronomy.
  1. Astronomy——Juvenile literature
  I. Title
  520   QB46

ISBN 0 00 191124 4

Printed in Great Britain by
Hazell Watson & Viney, Aylesbury

DICTIONARY OF SPACE and Astronomy

by ROBIN KERROD

GRANADA

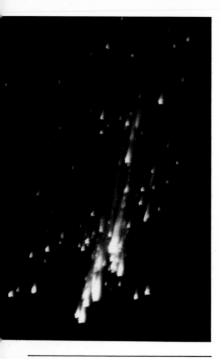

Ablation of its heat shield prevents a spacecraft burning up like this when it re-enters the atmosphere.

abort Cancel or cut short a space flight.

absolute magnitude The apparent magnitude a star would have if it could be observed at a distance of 10 parsecs (32.6 light-years). The absolute magnitude scale is a means of comparing star brightnesses.

absolute temperature A temperature scale whose starting point is the absolute zero of temperature. Temperatures are expressed in kelvins (K).

absolute zero The lowest possible temperature that can be reached, when all molecular motion ceases. Absolute zero is –273.16°C or 0K.

absorption spectrum A spectrum that shows a series of dark lines against a bright background. In the absorption spectrum of starlight the dark lines show where certain wavelengths have been absorbed by chemical elements in the star's outer atmosphere. The elements can be identified from the lines.

accelerometer A device for measuring acceleration. In inertial guidance systems three accelerometers are used to sense changes in acceleration along the three axes.

accretion The gathering of surrounding matter from space by a star or other body. The accretion by one star of gas from another is often thought to occur in binary systems. The gas spirals from one star to the other, forming a so-called accretion disc. There are thought to be accretion discs around black holes.

Achernar (Alpha Eridani) The ninth brightest star in the heavens, mag 0.48, in the constellation Eridanus. It is a hot blue star (B5), 127 light-years away.

achondrites A class of stony meteorites which do not show spherical

A

A-2 The code name for Russia's main launch vehicle, used for Soyuz launchings. It has two main rocket stages and four 'wrap-round' boosters. All rockets and boosters burn kerosene and liquid oxygen.

A-4 The original code name for the V-2 rocket.

aberration A defect of lenses and curved mirrors that causes a blurred image. See **chromatic aberration; spherical aberration.**

aberration of starlight The apparent displacement of stars due to the Earth's motion in space and to the finite velocity of light.

ablation Melting and boiling away, say of a heat shield during re-entry. This process dissipates the heat produced by air friction.

features (chondrules) in their structure.

achromatic lens A lens made from a combination of crown and flint glass in order to correct chromatic aberration, a defect that causes colour blurring of the image.

acquisition Locating a satellite or probe in space so as to collect tracking or telemetry data.

Acrux (Alpha Crucis) The 14th brightest star in the heavens (mag 0.9) in the constellation Crux. A hot blue star (B2), 260 light-years away.

Adams, John Couch (1819–1892) English astronomer who calculated the orbit of an eighth planet (Neptune) from perturbations in the orbit of Uranus. Leverrier also did so, independently.

Adhara (Epsilon Canis Majoris) A bright hot blue star (B2) in the constellation Canis Major, 22nd brightest in the heavens.

aerial Also called antenna; a rod, wire dish or array designed to pick up or transmit radio waves or microwaves.

A spectacular display of aerials at Carnarvon, Western Australia.

aerodynamic heating The heating that occurs because of air friction when a vehicle re-enters the Earth's atmosphere from space. See **communications blackout; heat shield.**

aerolite A stony meteorite. See **achrondrite; chrondrite.**

aeropause An indefinite region above the Earth where the atmosphere gradually merges into space.

aerospace A word coined from 'aeronautics' and 'space'. Aerospace vehicles are those that are able to operate in the atmosphere and also in space. The space shuttle can be considered the first true aerospace vehicle.

aether A hypothetical fluid that scientists once thought filled space. They thought that some kind of medium was necessary to carry light waves through space. But space is not completely empty; it contains scattered molecules of gas and dust. See **interstellar matter.**

age, of Moon The time that has passed since the last New Moon.

Agena A US rocket used in the 1960s as an upper stage in combination with an Atlas rocket for launching Ranger and Lunar Orbiter probes. It was also used as a docking target on the Gemini missions. The motor burned nitrogen tetroxide and hydrazine propellants.

air The mixture of gases that forms the atmosphere of the Earth. It is made up mainly of nitrogen (78%) oxygen (21%) and argon. Carbon dioxide, other gases, and water vapour are present in small quantities.

airglow The faint glow of the night sky caused by reactions between molecules of gas in the upper atmosphere.

airlock A chamber in a spacecraft from which the air can be removed without affecting the rest of the craft. Astronauts usually pass through an airlock when they go spacewalking.

Airpump See **Antlia.**

Al-Battani (AD 858–929) Also called Albetenius; the foremost Arab astronomer of the Middle Ages. He calculated more accurate values for the lengths of the year and the seasons and for the precession of the equinoxes.

albedo The extent to which a body reflects light falling on it. It is the ratio of the amount of light reflected to that received. The albedo of the Moon is only 0.07. Jupiter has the very high albedo of 0.7.

Alchor The star 80 Ursae Majoris. It forms a visual binary with the star Mizar (Zeta Ursae Majoris). They are the middle stars in the handle of the Plough.

Aldebaran (Alpha Tauri) The bright red giant star (K5) that forms the eye of the Bull in the constellation Taurus. At mag 0.86, it is the 13th brightest star in the sky. Around it is the Hyades open star cluster.

Aldrin, Edwin E. (born 1930) American astronaut who was the second person to walk on the Moon, on 20 July 1969, during the Apollo 11 mission.

Alfvén, Hannes (born 1908) Swedish astronomer who worked out a theory for the formation of the solar system using the effects of electro-magnetic forces.

Algol (Beta Persei) The first eclipsing binary star to be discovered, by John Goodricke in 1783. It varies in brightness from mag 2.2 to 3.5 every 2.9 days. It is sometimes called the 'Winking Demon'.

Left: An Agena target vehicle used for docking manoeuvres in orbit on the Gemini 11 mission in 1966.

Right: One of the most famous of all space pictures, Edwin Aldrin on the Moon in July 1969. It was taken by the first man on the Moon, Neil Armstrong, who is seen reflected in Aldrin's visor.

aliens Beings that live on other planets elsewhere in the universe. We do not know whether aliens exist, but it seems likely, considering the millions of suitable planets that ought to exist in other solar systems.

Almagest An encyclopedia of astronomy and mathematics written by Ptolemy of Alexandria in about AD 150. It gives details of the work of Greek astronomers such as Hipparchus. We know it in its Arabic translation. 'Almagest' means 'The Greatest'.

almanac A yearly publication containing expanded information about the calendar, listing astronomical information, religious events, bank holidays, etc. Specialist almanacs for astronomers and navigators include the *Astronomical Ephemeris*, formerly the *Nautical Almanac*.

Alpha Centauri Also called Rigil Kent; one of the nearest and brightest stars in the heavens, in the southern constellation Centaurus. Its magnitude is –0.2, and it lies only 4.3 light-years away. It is a multiple star, one of whose companions, Proxima Centauri, is slightly nearer to us.

Alpine Valley A prominent channel 130 km (80 miles) long that snakes through the lunar Alps.

Alps A mountain range on the Moon on the north-east of Mare Imbrium. It is cleft by Alpine Valley.

ALSEP An abbreviation for Apollo Lunar Science Experiments Package. It consisted of a selection of scientific instruments, such as seismometers and particle detectors, which the Apollo astronauts set up at their landing sites on the Moon. Powered by a nuclear generator, the instruments sent back readings automatically to Earth after the astronauts had left.

Altair (Alpha Aquilae) A hot white star (A7) in the constellation Aquila, 16 light-years away. At mag 0.77 it is the 11th brightest star in the night sky.

Altar See **Ara.**

altazimuth mounting A way of mounting a telescope so that it can move vertically – that is, in altitude, and horizontally – that is, in azimuth.

altitude The distance in degrees of a star above the horizon. It is one coordinate in the horizon system of stellar location.

Amalthea Jupiter's closest moon, which orbits at a distance of some 181,000 km (112,000 miles). It is only about 240 km (150 miles) in diameter.

Ames Research Center One of NASA's research centres, located near San Francisco. It manages spacecraft of the Pioneer series and is the base for the Kuiper Airborne Observatory. It also specializes in advanced flight technology.

Ananke One of Jupiter's tiny moons, less than 30 km (20 miles) across. It orbits at a distance of 20,700,000 km (12,800,000 miles).

Anders, William A (born 1933) One of the three US astronauts who took part in the first circumnavigation of the Moon in Apollo 8 in December 1968.

Andromeda A constellation in the northern hemisphere between Perseus and Pegasus. It is best known for its 'nebula', M31, which is actually a neighbouring galaxy.

Left: Three types of aliens are commonly reported. Some are dwarf-like with a large head; some are humanoid; others are hairy and ape-like.

Below: This picture of Alpine Valley was taken by a Lunar Orbiter probe. The Valley leads from Mare Imbrium through the lunar Alps.

The Andromeda galaxy.

Andromeda galaxy (M31,NGC224) Also called the Great Nebula in Andromeda; one of the few galaxies visible to the naked eye. It is a spiral galaxy like our own but twice as big, with a diameter of some 200,000 light-years. It lies about 2.2 million light-years away. M31 has a satellite galaxy M32. Both belong to the Local Group.

Andromedids See **Biela.**

Anglo-Australian Telescope A 3.9 m (154-inch) reflector at Siding Spring Observatory, New South Wales. Computer controlled, it is one of the most powerful instruments in the southern hemisphere. It was opened in 1974.

Ångstrom unit A unit used to measure the wavelength of light, named after the Swedish physicist Anders Jonas Ångström. $1\text{Å} = 10^{-10}$ metre.

angular distance Also called apparent distance. The distance between two heavenly bodies, expressed in angular measure, that is, degrees, minutes or seconds of arc.

Anik The name of a series of Canadian domestic communications satellites.

animals in space Animals were sent into space before humans, including mice, dogs (**Laika**) and monkeys. On the Skylab mission in 1973, one of the experiments was to observe a spider, 'Arabella', spinning a web.

annual parallax The displacement of the position of a star due to the Earth's annual motion around the Sun.

annular eclipse An eclipse of the Sun in which the Moon does not quite cover the disc of the Sun. You see a bright ring, or annulus, around the dark Moon.

Antares (Alpha Scorpii) A red supergiant star (M1) whose diameter is some 400 times that of the Sun.

antenna Another term for an aerial.

antimatter A kind of matter made up of atomic particles that are opposites to ordinary atomic particles. Thus, an ordinary electron has a negative electric charge. Its antiparticle, a positron, has a positive charge. An antiproton has a negative charge.

Antlia The Airpump; an inconspicuous southern constellation.

Antoniadi, Eugène Michael (1870–1944) French astronomer who devised the modern scale of seeing.

Apennines A prominent mountain range on the Moon, bordering Mare Imbrium. It lies between the craters Archimedes and Eratosthenes, rising in places to some 6000 m (20,000 ft).

aperture The clear diameter of a lens or mirror.

aperture synthesis A method of creating a large-aperture radio telescope dish. It uses the output from

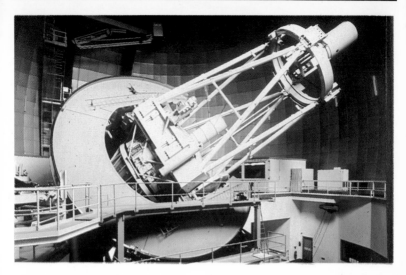

Above: The Anglo-Australian Telescope is typical of the new generation of reflectors, light in weight and under computer control.

Below: Edwin Aldrin steps down on to the Moon during the first Apollo landing, July 1969.

a number of smaller dishes placed different distances apart. The large 'dish' is created as the small dishes move round because of the Earth's rotation. Martin Ryle developed this technique at Cambridge.

aphelion The point in an orbit of, say, a planet, that is farthest from the Sun. At aphelion, the Earth lies 152,000,000 km (94,450,000 miles) from the Sun.

Aphrodite Terra The largest of the two main highlands on Venus, found by radar scanning from space probes in orbit. It covers an area half the size of Africa near the planet's equator.

apogee The point in the orbit of the Moon or an artificial satellite that is farthest from the Earth. The Moon at apogee lies some 406,000 km (252,000 miles) away.

Apollo landings Six lunar landings took place. Apollo 11 set down on

An Apollo lunar module on the Moon. After exploration, the astronauts returned to orbit in the upper part of the craft.

Mare Tranquillitatis on 20 July 1969. First man on the Moon was **Neil Armstrong.** Apollo 12 went to the Ocean of Storms. (Apollo 13 was aborted when the spacecraft was crippled on the outward journey. The crew returned safely after using the lunar module as a 'life-raft'.) Apollo 14 explored a region near Fra Mauro crater. Apollo 15 visited the foothills of the Apennines near Hadley Rille. Apollo 16 landed in the Descartes Highlands. Apollo 17 explored the Taurus-Littrow valley.

Apollo project A US Moon landing project that landed 12 astronauts on the Moon between July 1969 and December 1972. The Apollo spacecraft consisted of three modules. The command module carried the crew of three. The service module carried the main equipment. And the lunar module was the part that carried

two astronauts down to the lunar surface. The whole spacecraft was lofted into space by the gigantic Saturn V rocket. Of the Apollo flights, Apollo 7 was the first manned mission, in Earth orbit. Apollo 8 circumnavigated the Moon but did not land. Apollo 9 tested the whole Apollo spacecraft in Earth orbit. Apollo 10 tested the whole Apollo craft in lunar orbit. Apollo 11 was the first Moon landing. Apollo 17 was the last.

Apollo-Soyuz Test Project (ASTP) The first joint US/Russian manned space flight, which took place in July 1975. Three US astronauts in an Apollo spacecraft took part in experiments with two Russian cosmonauts in a Soyuz craft, while the two craft were linked together by a specially designed docking module.

apparent magnitude The magnitude of a star as observed from the Earth. It seldom bears any relation to its true, or absolute magnitude.

apsides The two points in the orbit of a planet or satellite comet when it is nearest or farthest away from its parent body, eg the Sun. Perihelion, for example, is the apside of the Earth when it is closest to the Sun.

Apus The Bird of Paradise; a faint constellation in the southern hemisphere, quite close to the South Pole.

Aquarids Two meteor showers with their radiant in the constellation Aquarius. The Eta Aquarids are seen in the first week in May, the Delta Aquarids at the end of July.

Aquarius The Water-Bearer; a large constellation of the zodiac, close to the celestial equator. It has no really bright stars. Perhaps its most striking feature is the magnificent globular cluster M2.

The Apollo command and service modules (CSM), measuring about 10.5 m (34 ft) long. The forward command module was the only part to return to Earth.

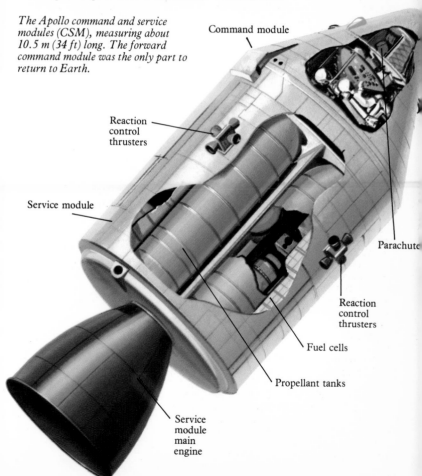

Command module

Reaction control thrusters

Service module

Parachute

Reaction control thrusters

Fuel cells

Propellant tanks

Service module main engine

Aquila The Eagle; a fine constellation on the celestial equator. Its brightest star is the first magnitude **Altair,** one of the stars of the **summer triangle.**

Ara The Altar; a constellation in the southern hemisphere.

Archer See **Sagittarius.**

Arecibo radio telescope A huge radio telescope at Arecibo on the island of Puerto Rico. Its dish, 305 m (1000 ft) across, is suspended in a natural bowl. The dish is made up of 38,778 perforated aluminium panels.

Ariane The main launch vehicle of the European Space Agency, first flight tested in 1979. Some 47 metres (155 ft) long, it is a three-stage vehicle.

The first two stages burn hydrazine (fuel) and nitrogen tetroxide (oxidizer). The third stage burns liquid hydrogen and liquid oxygen. Ariane 3 has twin solid rocket boosters.

Ariel One of the moons of Uranus. Some 850 km (550 miles) in diameter, it orbits 192,000 km (120,000 miles) from the planet.

Ariel (Spacecraft) A series of British scientific satellites launched by NASA. Ariel 5 was a pioneer of X-ray astronomy.

Aries The Ram; a zodiacal constellation of little interest that lies between Taurus and Pisces. See also **First point of Aries.**

Below: The giant Arecibo radio telescope in Puerto Rico has beamed messages to other stars.

Right: Ariane stands on the launch pad at the Guiana Space Centre, Kourou, French Guiana.

Aristarchus of Samos (310–230 BC) Greek philosopher who first advanced the idea that the Earth circles the Sun and not vice versa. He also estimated the relative sizes of the Moon and the Earth, quite accurately for the time.

Aristotle (384–322 BC) Greek philosopher who favoured an Earth-centred universe in which the heavenly bodies rotated on concentric spheres around a spherical Earth.

Arizona meteor crater The biggest metereorite crater on Earth, located near Winslow in Arizona. Some 1265 m (4150 ft) across, it is 175 m (575 ft) deep. It is thought to be about 50,000 years old.

Armstrong, Neil (born 1930) American astronaut who became, on 20 July 1969, the first human being to set foot on the Moon. As he stepped onto the surface from the Apollo 11 lunar module he said: 'That's one small step for a man, one giant leap for mankind'.

array A set of aerials or antennae.

Arrow See **Sagitta.**

artificial gravity A kind of gravity created by rotating a spacecraft. The centrifugal forces set up mimic gravity. Space colonies of the future may have artifical gravity.

artificial satellite Any spacecraft launched into orbit around the Earth. Usually the word 'artificial' is omitted. Sputnik 1 was the world's first artificial satellite when it went into orbit on 4 October 1957.

ASAT Short for anti-satellite satellite. It is a spacecraft able to destroy other satellites by means of explosion, radiation or laser beams. It is a so-called 'Star Wars' weapon.

ashen light A faint glow observed on the dark side of the crescent Venus. It is probably an aurora-like glow caused by disturbances in the planet's upper atmosphere. See also **earthshine**.

In the shuttle astronauts often fly into space seven at a time.

aspect The position of a heavenly body relative to another, eg of a planet relative to the Sun. See **conjunction; opposition; quadrature.**

associations Loose groups of mainly young stars extending over a distance of up to several hundred light-years. O-associations consist of O and B type stars; T-associations consist of young red dwarf stars.

A stars Hot white stars of spectral type A, with a temperature of some 10,000K.

asterism A small group of stars, not necessarily a constellation.

asteroid Also called minor planet and planetoid; a small body orbiting the Sun. Most asteroids orbit in a ring, or 'belt', between the orbits of Mars and Jupiter, on average about 450,000,000 km (280,000,000 miles) from the Sun. Some, such as Eros and Icarus, stray quite close to the Earth's orbit.

ASTP
See **Apollo–Soyuz Test Project.**

astrolabe An instrument used by ancient astronomers and navigators to measure the angle of a heavenly body above the horizon. At its simplest it was a graduated circular disc with a movable sighting arm.

astrology A pseudo-science in which the heavenly bodies are supposed to influence human affairs. Astrology grew up in Chaldea in the Middle East about 3000 years ago. In the stargazing of the ancient astrologers can be seen the birth of the true science of astronomy.

astrometry The branch of astronomy concerned with measuring the positions and motions of the heavenly bodies. It is often called positional astronomy.

astronaut A space traveller; literally a traveller to the stars. The Russian term is cosmonaut.

astronautics The science and technology of space flight. Konstantin Tsiolkovsky is often considered the 'father of astronautics'.

Astronomer Royal An honorary title awarded to one of Britain's top astronomers. The post was created by Charles II in 1675. John Flamsteed was the first Astronomer Royal. Until 1972 the Astronomer Royal was also

Director of the Royal Greenwich Observatory. Now the two posts are separate.

astronomical coordinates
A system of pinpointing a heavenly body on the celestial sphere. In the horizon system the coordinates are altitude and azimuth. In the equator system they are right ascension and declination.

astronomical unit This is the mean distance between the Earth and the Sun, some 150 million km (93 million miles).

astronomy The scientific study of the heavens and of all the heavenly bodies – stars, planets, comets, clusters, galaxies, nebulae, quasars, pulsars, black holes, etc. Astronomy has its roots in the naked-eye star-gazing of the astrologers of early civilizations in the Middle East. Galileo pioneered telescopic observations in 1609. Karl Jansky founded radio astronomy in 1931. Space astronomy became possible after the launching of Sputnik 1 in 1957. Today astronomers can peer at the heavens in nearly all wavelengths of the electromagnetic spectrum.

The dome housing the world's largest reflector, the 6-m (236-inch) telescope at Zelenchukskaya in Russia.

The giant stone sundial at the observatory of Jai Singh at Jaipur in India. The observatory was set up in 1734.

astronomy satellite A spacecraft that carries instruments and cameras into orbit. Because it travels above the Earth's atmosphere, it can look at the heavens at wavelengths (X-ray, gamma-ray, infrared) that are absorbed by the Earth's atmosphere and therefore cannot be studied from the ground.

astrophotography Photographing the heavens. Most big telescopes are used as cameras these days to photograph the stars. This is preferred to direct observation because film can store the light that falls on it. Long exposures will therefore show up faint objects.

astrophysics The branch of astronomy concerned with the physical properties and composition of the heavenly bodies.

Atkov, Oleg (born 1949) Russian cosmonaut who was one of the three-man crew that established a new space endurance record of 237 days in orbit in Salyut 7. They returned to Earth on 2 October 1984.

Atlantis The name of the fourth space shuttle orbiter, which became operational in 1985.

Atlas A US rocket used to launch the first astronauts into space in the Mercury project. Some 21 m (69 ft) long, it used kerosene and liquid oxygen as propellants. Combined with Agena and Centaur rockets, it was used to launch early lunar probes. The Atlas-Centaur is still used to launch geostationary satellites, such as Intelsat V.

atmosphere The layer of gases around a planet. The Earth has an atmosphere of air, mainly nitrogen and oxygen. Other planets have thick atmospheres, including Venus, Jupiter, Saturn, Uranus and

Neptune. Mars has a very thin atmosphere. Mars and Venus have atmospheres of mainly carbon dioxide. Those of the other planets are mainly hydrogen and helium.

ATS-6 A highly successful communications satellite that operated from 1974 to 1985. In geostationary orbit, it was used for linking remote communities in North America and later for educational broadcasting to Indian towns and villages.

AU The abbreviation for **astronomical unit.**

Auriga The Charioteer; a prominent constellation in the northern hemisphere, partly in the Milky Way. Shaped like a kite, its chief star is the magnificent mag 1 Capella.

aurora Waves and folds of coloured light that can be seen in far northern and far southern skies. It is caused by the collision between molecules in the upper air and streams of particles coming from the Sun. In north polar regions it is called the aurora borealis, or northern lights; in the south, aurora australis, or southern lights.

australite A kind of **tektite** found in australia.

autumnal equinox The time of the year when the Sun is exactly over the equator, moving south, and day and night are of equal length all over the world. It occurs on about September 23. It marks the beginning of autumn in the northern hemisphere, and of spring in the southern.

azimuth One of the coordinates in the horizon system of locating stars. It is the horizontal angle between the vertical circle through the star and the plane of the meridian of the observer.

Left: Three successful astronomy satellites: top left, an Orbiting Astronomical Observatory; top right, Ariel 4; bottom, a High Energy Astronomical Observatory.

Below: A display of the aurora borealis, photographed near Kiruna in Sweden. It takes the form of a shimmering curtain of coloured light.

B

Baade, Wilhelm Heinrich Walter (1893–1960) German-born astronomer who worked at Mt Wilson and Palomar Observatories from the 1930s. He is noted for his theories of star populations.

background radiation A faint microwave radiation that permeates the universe. It has a temperature of about 3K. It is thought to be the radiation left over from the Big Bang that created the universe. Also called fireball radiation, it was discovered in 1965 by Arno Penzias and Robert Wilson.

backup A system in a spacecraft that can take over the work of the prime system if that fails. Such systems are called redundant. The term is also supplied to astronauts who can replace the prime crew on a space mission.

Below: Soyuz T-11 lifts off from Baikonur Cosmodrome, April 1984.

Baikonur Cosmodrome The main Russian launch centre, near the town of Tyuratam in southern Russia to the west of the Aral Sea. It covers an area about 140 km (87 miles) long and 90 km (56 miles) wide and is thus very much bigger than the Kennedy Space Center. It is the launch site for all the manned Soyuz missions to the Salyut space stations.

Baily's beads Bright beads of light that can be seen around the limb of the Moon during a total solar eclipse. The effect is caused by sunlight shining through valleys on the Moon. It is named after the person who first observed it, the English astronomer Francis Baily.

Barnard, Edward Emerson (1857–1923) American astronomer who discovered many bodies, including the dwarf star now named after him.

Barnard's star A red dwarf star in the constellation Ophiuchus, which was discovered by E. E. Barnard in 1916. It has the largest proper motion of any star, some 10.3 seconds of arc a year. It is thought that it could have one or more large planets circling around it.

barred-spiral galaxy A major class of galaxy, notable for a bar across its nucleus.

Barringer crater Another name for the **Arizona meteor crater.**

barycentre The centre of mass, in particular that of the Earth-Moon system. The barycentre of this system actually lies within the Earth.

beam-builder A device developed by NASA for automatically constructing aluminium trussed girders, or beams. Designed to be carried in the space shuttle, the beam-builder will in the future be used to manufacture beams in orbit for use in space station construction.

Beehive See **Praesepe.**

Above: A barred-spiral galaxy.

A beam-builder, which makes aluminium beams automatically.

Bellatrix (Gamma Orionis) A 1st magnitude star that marks the left shoulder of the figure of Orion. It is a blue giant (B2) 300 light-years away.

Berenice's Hair
See **Coma Berenices.**

Bessel, Friedrich Wilhelm (1784–1846) German astronomer, who in 1838 first calculated the distance to a star (61 Cygni) from measurements of its parallax.

Beta Centauri See **Hadar.**

Beta Lyrae stars Eclipsing variable stars named after the prototype, Beta Lyrae. They consist of two stars revolving close together which are distorted and surrounded by a gaseous cloud.

Betelgeuse (Alpha Orionis) A red supergiant star (M2) in the constellation Orion. At mag 0.8, it is the 12th brightest star in the sky. It lies some 650 light-years away. It is the only star besides the Sun to have had its surface photographed, or rather electronically imaged.

Right: A cluster of galaxies in Hercules, rushing away from us.

Astronomers think that a colossal explosion (Big Bang) created the universe 15–20,000 million years ago and set it expanding.

Biela, Wilhelm von (1782–1856) Austrian astronomer who discovered the short-period comet named after him in 1826. Biela's comet, with a period of 6.6 years, split into two in 1846, was seen in 1852 but not since. In its place now is a regular meteor shower, the Andromedids.

Big-Bang theory A theory about the origin of the universe now favoured by most astronomers. According to this theory, the universe, space and time began about 15,000–20,000 million years ago when an intensely hot ball of matter (a fireball or primeval atom) exploded and began expanding. As the expansion continued, the temperature of the universe fell rapidly. Matter in the form of gas began condensing into stars and galaxies. The expansion of the universe continued, and still continues today (see **expanding universe**). Evidence for a Big Bang was provided by the discovery of the general background radiation that now permeates the universe.

Big Bird A huge US spy satellite, about 15 m (50 ft) long and weighing some 11 tonnes. Circling in a relatively low orbit (down to 160 km, 100 miles), Big Bird carries high-resolution cameras and ejects film in capsules, which are recovered from the air. It can spot people on the ground.

Big Crunch The name given to an event that could happen in a closed universe, when the expansion of the universe is halted and contraction begins. Eventually all matter will come together in a Big Crunch, the reverse of the Big Bang.

Big Dipper The American name for the **Plough.**

binary star A double-star system in which the two stars are physically associated. This contrasts with an optical double, in which the stars appear to be close together but are actually far apart in space. In true binary systems both stars revolve around a common centre of mass. The two components of a binary may be sometimes separated visually (visual

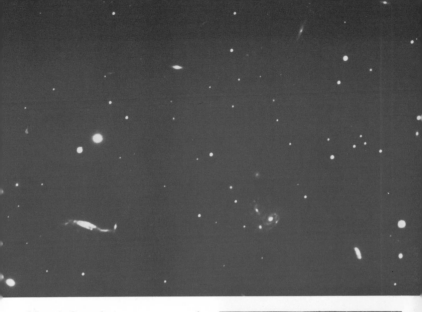

binary). Sometimes a spectroscope is required (spectroscopic binary). Others are identified by their light curve (eclipsing binary).

binoculars A kind of compact two-tube telescope. Prismatic binoculars use combinations of prisms to 'fold' the path of light between the objective and eyepiece lenses. For general purposes 10 × 50 binoculars (10 times magnification, 50 mm diameter objectives) are suitable. For serious comet and nova spotting 20 × 80 binoculars are better.

bird A slang term for a rocket or a satellite.

Bird of Paradise See **Apus**.

Black Arrow A British three-stage launch vehicle that put the satellite Prospero into orbit in October 1971. With this launch Britain became the sixth nation to launch a satellite independently.

Right: The Big Crunch theory suggests that one day the universe will start to shrink and come together into a dense mass.

black hole A region of space where gravity is so enormous that nothing, not even light, can escape from it. Black holes are thought to result from the collapse of supermassive stars. The boundary of a black hole is termed the event horizon. Black holes probably exist at the centres of some galaxies. According to current thinking, black holes are part of binary star systems. An accretion disc of gas spiralling from the outer star into a black hole would give rise to powerful X-rays. It is by such X-rays that black holes would be identified. The strong X-ray source Cygnus X-1 might well be a black hole.

blackout A condition in which a person is rendered unconscious when blood rushes from the brain as a result of extreme acceleration, or high G-forces. This can occur when a pilot pulls out of a high-speed dive. Steps are taken in space flight to limit the G-forces at lift-off and during re-entry to prevent the crew blacking out. (See also **communications blackout**.)

blink microscope Also called comparator, an instrument astronomers use to detect differences in two photographs taken of the same part of the sky. The photographic plates are presented to the viewer in rapid succession. Any change in star position or brightness shows up.

blue shift This is the shift of the spectral lines of starlight towards the blue end of the spectrum. It indicates that the star is approaching us. See also **red shift**.

Bode, Johann Elert (1747–1826) German astronomer who helped derive the strange relationship between the distances of the planets, now known as Bode's Law.

Bode's law Or Titius-Bode law; an empirical law which gives the distances of the planets from the Sun. If a is the mean distance to the Sun in astronomical units, then $a = 0.4 + 0.3 \times 2^n$, where n = infinity, 0,1,2 etc. The law gives a figure for a in close agreement with the actual figures for all the planets out to Uranus. No one knows why this should be.

Bok globule A spherical dark nebula only a few light-years across. Several are known in different parts of the sky. It is named after the Dutch-American astronomer Bart Bok.

bolide Also called a fireball; an exceptionally bright meteor, which sometimes explodes. It often gives rise to meteorites.

bolometer A sensitive instrument for measuring radiation. A common type contains a platinum wire, whose resistance changes when radiation falls on it. The bolometric magnitude is a measure of the total radiation given out by a star.

Bondi, Hermann (born 1919) Austrian astronomer working in England, who was one of the authors (with Hoyle and Gold) of the steady-state theory of the universe.

booster A rocket that is attached to the main launch vehicle to provide added take-off thrust. The space shuttle uses two large solid rocket boosters.

Left: Illustration of what could happen around a black hole. Gas spirals into the black hole from a companion star, heats up and gives out X-rays.

Right: Nine booster rockets help blast the Delta launch vehicle off the launch pad.

Boötes The Herdsman; a prominent northern constellation close to the Northern Cross. Its major star is the brilliant Arcturus, fourth brightest star in the heavens.

Borman, Frank (born 1928) One of three US astronauts who took part in the first circumnavigation of the Moon in Apollo 8 in December 1968.

Bradley, James (1693–1762) English astronomer who became Astronomer Royal in 1742. He is noted for his discovery of the aberration of starlight and for compiling an accurate star catalogue.

Brahe, Tycho (1546–1601) Danish astronomer who was a brilliant observer of the heavens. His observations helped his one-time assistant Kepler to formulate his laws of planetary motion. In 1572 he saw a supernova, often called Tycho's star.

breccia A common type of rock found on the Moon, made up of cemented rock chips.

brightness The intensity of starlight. Brightnesses are assessed on a scale of **magnitude.**

British Astronomical Association (BAA) Britain's foremost astronomical society for amateur observers, founded in 1890. Its publications include a monthly *Journal* and an annual *Handbook.*

British Interplanetary Society (BIS) One of the world's earliest societies devoted to the study of astronautics, founded in 1933. Amongst its publications are the magazines *Spaceflight* and *Space Education,* and the *Journal of the BIS.*

B stars Hot blue-white stars of spectral type B, which have a high absolute magnitude. Their surface temperature is about 20,000°C.

Bull See **Taurus.**

burn The period when a rocket engine is firing.

burster An object that gives out machine-gun-like bursts of X-rays of enormous energy. Bursters are

probably neutron stars that undergo nuclear explosions when gas builds up on their surface.

butterfly diagram A graph showing the drift of sunspots towards the solar equator during the 11-year solar cycle. It shows a characteristic pattern of butterfly wings.

C

3C–273 The first quasar to be discovered (number 273 in the third Cambridge Catalogue of radio sources). In 1962 astronomers found that the radio source 3C–273 coincided with a star-like body. This body was found to be over 2000 million light-years away and so had to be hundreds of times brighter than an average galaxy. Yet it was only a fraction of the size of a galaxy. It was called a quasi-stellar object, or **quasar.** Many hundreds are now known.

Caelum The Chisel; a tiny southern constellation.

calendar The division of the year into months, weeks and days. Calendars have been devised since the beginning of civilization for the convenience of society to allow the regular celebration of religious events, the planting and harvesting of crops, etc. The present calendar is a solar calendar, based on the 365¼ days it takes the Earth to orbit the Sun. It is called the Gregorian calendar, after Pope Gregory XIII, who introduced it in 1582. See also **Gregorian calendar; Julian calendar; leap year; lunar calendar.**

Left: A sample of lunar breccia, a rock made up of cemented chips and crystals.
Below: The BIS's 1948 design for a space station.

Below: The first quasar found, 3C–273. It is notable for its incredibly high energy output and a peculiar jet that projects from its centre.

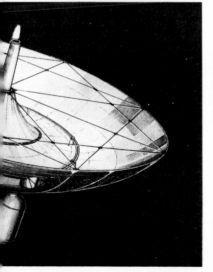

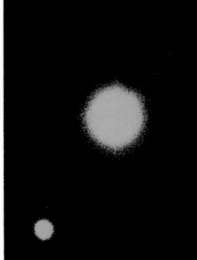

Above: Jupiter's moon Callisto, pictured by the Voyager 1 space probe in 1979.

Callisto The second largest of Jupiter's moons, 4820 km (2995 miles) across. It orbits at a distance of 1,900,000 km (1,200,000 miles). Discovered by Galileo in 1610, it has an ancient, heavily cratered surface.

Caloris Basin An enormous impact basin on Mercury, some 1300 km (800 miles) in diameter.

Camelopardalis The Giraffe; a faint, long and straggling northern constellation between Ursa Major and Cassiopeia.

canals Linear features on the surface of Mars reported in 1877 by Giovanni Schiaparelli and later by other astronomers, notably Percival Lowell. People assumed the 'canals' to be artificial and imagined there to be a race of intelligent Martians. Space probes have now confirmed that there are no canals or Martians. There are natural channels, but these are not visible from Earth.

Astronomer Percival Lowell drew this map of Martian 'canals' in the late 19th century, convinced that they were dug by Martians.

Cancer The Crab; a faint constellation of the zodiac between Leo and Gemini. Its most interesting feature is the fine open star cluster Praesepe, or the Beehive (M44).

Canes Venatici The Hunting Dogs; a faint northern constellation that lies beneath the handle of the Plough. Its chief star, known as Cor Caroli (Charles' Heart), is an optical double.

Canis Major The Great Dog; a southern constellation near the celestial equator, easily located because of its main star Sirius, the Dog Star, which is the brightest in the heavens. It also contains another 1st magnitude star, Adhara, and a rich star cluster (M41).

Canis Minor The Little Dog; a small constellation located near the celestial equator close to the Milky Way. It is dominated by the brilliant Procyon, eighth brightest star in the sky.

Cannon, Annie Jump (1863–1941) American astronomer noted for her classification of stellar spectra and compilation of the Draper Catalogue.

Canopus (Alpha Carinae) A brilliant supergiant star (F0) in the southern constellation Carina. At mag –0.73, it is the second brightest star in the heavens. It lies 196 light-years away.

Cape Canaveral A cape on the east coast of Florida, USA, from which US rockets are launched. Just inland from the Cape, on Merritt Island, is the Kennedy Space Center, the prime launch and landing site for the space shuttle.

Capella (Alpha Aurigae) At mag 0.08, giant star Capella (G8) is the sixth brightest star in the sky, in the constellation Auriga. It is a spectroscopic binary with a period of 105 days. It lies 46 light-years away.

Capricornus The Sea Goat; a faint zodiacal constellation between Aquarius and Sagittarius.

capsule The small pressurized crew cabin of early spacecraft like Mercury.

captured rotation This is also called synchronous rotation; a condition in which a satellite spins on its axis in exactly the same time as it takes to orbit its parent plant. Because of this, the satellite always keeps the same face towards the planet. The Moon has a captured rotation around the Earth.

The Mercury capsule, in which US astronauts first flew into space. Inside, there was virtually no room to move.

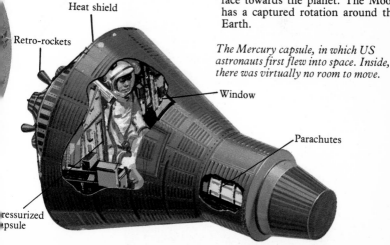

Heat shield

Retro-rockets

Window

Parachutes

Pressurized capsule

carbon-nitrogen cycle One type of nuclear reaction thought to take place in the interior of stars that provides energy to keep the stars shining. In the reaction carbon acts as a catalyst to make four protons combine into a helium atom. Nitrogen is formed during the reaction.

Carina The Keel; a large sprawling far southern constellation. It extends from the brilliant Canopus nearly to Crux, the Southern Cross.

Carme One of Jupiter's tiny outer moons. On average about 22 million km (14 million miles) from the planet, its diameter can be no more than 40 km (25 miles).

Carpathians A rugged mountain range on the Moon, on the south-west edge of Mare Imbrium, close to the crater Copernicus. They rise to some 2100 m (7000 ft).

Cassegrain focus A very common arrangement for viewing the image in a reflecting telescope. Light is gathered by a curved primary mirror and reflected up to a small convex secondary mirror. This in turn reflects the light back down the telescope tube through a hole in the centre of the primary mirror. The light is then brought to a focus conveniently behind the mirror.

Cassini division A dark gap between Saturn's A and B rings, first observed by Giovanni Cassini in 1675. It measures some 4000 km (2500 miles) wide.

Cassini, Giovanni Domenico (1625–1714) Italian astronomer who

Right: In this close-up picture of Saturn's rings, the Cassini division shows up clearly. Also in this picture can be seen two of the tiny shepherd moons orbiting around the outer edge of the rings.

Left: In Carina is a spectacular nebula around the peculiar star Eta Carinae. The nebula is crossed by dark lanes of obscuring opaque material.

discovered four of Saturn's moons and the dark gap in Saturn's rings.

Cassiopeia One of the most beautiful northern constellations, with an unmistakable W-shape. Its five main stars are of the 2nd and 3rd magnitude.

Cassiopeia A The most powerful radio source in the sky, in the constellation Cassiopeia. It is thought to be the remnant of a supernova that occurred in the 1600s.

Cassiopeia B Another powerful radio source in Cassiopeia, which is the remnant of the supernova we know as Tycho's star, witnessed by Tycho Brahe in 1572.

Castor (Alpha Geminorum) The second brightest star in the constellation Gemini. It is the twin star of Pollux. Castor is a 1st mag multiple star which has at least six components.

Caucasus Mountains A high lunar mountain range between Mare Serenitatis and Mare Imbrium. Some peaks are more than 6000 m (20,000 ft) high.

CCD Abbreviation for **charge-coupled device.**

celestial coordinates
See **astronomical coordinates.**

celestial mechanics A branch of astronomy concerned with the motion of the heavenly bodies.

celestial sphere An imaginary sphere around the Earth to which the stars seem fixed. Astronomers use this ancient concept to provide a means of reference for locating the stars. See **astronomical coordinates.** The celestial poles are located directly above the Earth's poles. The celestial equator is the projection on the celestial sphere of the Earth's equator. Because the Earth revolves on its axis from West to East, the celestial sphere appears to revolve from East to West.

Centaur See **Centaurus.**

Centaur Rocket used as an upper

stage in the US heavy launch vehicles Atlas-Centaur and Titan-Centaur. The latter is used for launching deep space probes. The Centaur burns liquid hydrogen and liquid oxygen propellants and develops some 14 tonnes of thrust.

Centaurus The Centaur; one of the most magnificent of the constellations, in far southern skies. It contains the 3rd brightest star in the sky, Alpha Centauri, and the 10th brightest, Hadar, or Beta Centauri. Another unmistakable feature of the constellation is the brilliant naked-eye star cluster Omega Centauri.

Centaurus A The third most powerful radio source in the sky, in the constellation Centaurus. In photographs it looks like a spherical galaxy cut in two by a dark band. It pours out enormous energy at visible and invisible wavelengths, indicating violent activity inside it.

centrifuge A machine in which astronauts sit in a gondola and are whirled round at the end of a long arm. In the gondola the astronauts experience centrifugal forces that mimic the G-forces they will experience during launch and re-entry.

Cepheid A variable star that changes in brightness with absolute precision over a precise period. Cepheids are named after the prototype, Delta Cephei, whose magnitude varies between 3.8 and 4.6 in $5\frac{1}{3}$ days. Delta Cephei is one of the long-period, or classical Cepheids, whose periods can

A peculiar galaxy in Centaurus, which pours out enormous energy. Known as Centaurus A, it lies about 16 million light-years away. It is the nearest of the so-called active galaxies. It is roughly spherical in shape, measuring about 150,000 light-years across.

range from 1 to 50 days. Cepheids with shorter periods are usually classed as RR Lyrae stars. Cepheids are useful for determining astronomical distances, see **period-luminosity law.**

Cepheus A northern constellation close to Cassiopeia. Its most interesting star is Delta Cephei, which is the prototype for the class of variable stars known as the Cepheids.

Ceres The largest asteroid, measuring some 1000 km (600 miles) across. It was discovered by Giuseppe Piazzi on 1 January 1801.

Cerro Tololo Inter-American Observatory One of the foremost observatories in the southern hemisphere. It is located 2200 m (7200 ft) up in the Andes near La Serena in Chile, some 500 km (300 miles) from Santiago. Its major instrument is a 4-m (157-in) reflector.

CETI An abbreviation for 'communication with extraterrestrial intelligence'.

Cetus The Sea Monster; a sprawling constellation near the celestial equator. Its most notable star is the eclipsing binary Algol, the 'Winking Demon'.

Chaffee, Roger B. (1935–1967) US astronaut, who was one of three killed when fire gutted an Apollo training capsule.

Challenger The second operational space shuttle orbiter, which made its space début in June 1983.

Chamaeleon An inconspicuous southern constellation.

Chandrasekhar, Subrahmanyan (born 1910) Indian-born US astrophysicist who works on the evolution of stars.

chaotic terrain Some regions of jumbled, fractured rock found on certain parts of Mars, from which channels often emerge. They could have been caused by the rapid thaw of sub-surface ice.

Left: A charge-coupled device, now widely used by astronomers. The light-sensitive chip is mounted in the middle.

Below: The Viking 1 probe took this picture of Chryse when it landed on Mars in 1976. The ground is rusty red in colour and strewn with rocks. Here and there are small sand dunes.

charge-coupled device (CCD) A large light-sensitive silicon chip, used as a light detector in telescopes. CCDs provide an electronic image, which is then fed into a computer. The computer can then display a visual image in a variety of false colours to bring out particular detail. The latest CCDs have hundreds of thousands of light-sensitive picture elements, or pixels.

Charioteer See **Auriga.**

Charon The moon of Pluto, discovered in 1978. It is about 800 km (500 miles) in diameter.

chemical rocket A conventional rocket in which chemical reactions, usually burning, produce the hot gases for propulsion.

Chiron A peculiar faint member of the solar system discovered by C.T. Kowal in 1977. It could be an asteroid or an escaped moon of Saturn.

Chisel See **Caelum.**

chondrite A kind of meteorite which contains spherical features called chondrules. Some contain varying amounts of organic, or carbon-based matter, and are termed carbonaceous chondrites.

Chrétien, Jean-Loup (born 1938) The first West European to go into space, as a guest cosmonaut on the Soyuz mission to the Salyut 6 space station in June 1982.

chromatic aberration A defect of lenses which results in the colour

blurring of the image. It is caused by light of different wavelengths (colours) being brought to a focus in slightly different places. See **achromatic lens.**

chromosphere The lower layer of the Sun's atmosphere. From Earth it is visible as a reddish band only at times of total solar eclipse.

Chryse A plains region on Mars (Chryse Planitia) on which the Viking 1 lander set down in July 1976.

Circinus The Compasses; a small southern constellation, near Centaurus.

circumpolar stars Stars that are always above the horizon from the observer's latitude. From Britain, for example, the prominent constellations Ursa Major and Cassiopeia are circumpolar.

Clarke orbit Another term for geostationary orbit, named after the noted space writer Arthur C. Clarke. He was first to suggest this orbit as a prime location for communications satellites in an article he wrote for *Wireless World* in 1945.

Clavius The largest lunar crater, measuring over 230 km (140 miles) across, located near the southern limb of the Moon.

Clock See **Horologium.**

Clock drive A means of rotating an equatorially mounted telescope around the polar axis to compensate for the spin of the celestial sphere. In this way a star under observation remains in the same spot in the telescope.

closed universe A concept for the evolution of the universe, which suggests that the present expansion will eventually cease (see **expanding universe**). Then the universe will start to contract, ending with a 'Big Crunch' as all matter comes together.

cluster, star See **globular cluster; open cluster.**

clusters, of galaxies Groups of galaxies that exist in space. The largest, containing hundreds of galaxies, occur in the Coma Berenices-Virgo region.

cluster variable An alternative name for **RR Lyrae stars.**

Left and below: A comet grows a tail as it approaches the Sun. The tail always points away from the Sun.

Opposite: One of the Apollo command modules that came back from the Moon.

Coal Sack A conspicuous dark nebula of dust and gas silhouetted against the Milky Way in the far southern constellation Crux.

collapsar An alternative name for **black hole.**

collimation The process of aligning the optical axis of a reflecting telescope, essential before observations begin.

Collins, Michael (born 1930) US astronaut who piloted the Apollo 11 command module in July 1969, remaining in lunar orbit while his fellow astronauts Armstrong and Aldrin landed on the Moon.

Colt See **Equuleus.**

Columba The Dove; a small southern constellation.

Columbia Name of the first operational space shuttle orbiter, which made its maiden flight on 12 April 1981, crewed by veteran astronaut John Young, and Robert Crippen.

coma The cloud of gas and dust around the nucleus of a comet.

Coma Berenices Berenice's Hair; a northern constellation that demands interest, not because of its individual stars but because of its rich field of globular clusters and galaxies (see below).

Coma cluster A cluster of galaxies in the constellation Coma Berenices, containing about 800 members. It lies 350 million light-years away.

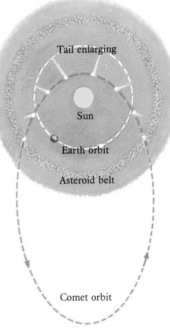

Tail enlarging

Sun

Earth orbit

Asteroid belt

Comet orbit

comes The least bright component of a binary star system.

comet A tiny member of the solar system, which becomes visible only when it approaches the Sun. Composed of dust and ice, it is often described as a 'dirty snowball'. It starts to shine when solar heat evaporates off gas and releases dust, which then reflects sunlight. The pressure of the solar radiation pushes the dust away from the comet's head, creating a visible tail that always points away from the Sun. Comets often appear without any warning, but a few return to the vicinity of the Sun regularly. See **Encke's comet; Halley's comet; Oort cloud.**

command module The cone-shaped section of the Apollo spacecraft that carried the crew and the only part to return to Earth after a mission.

companion of Sirius The faint component of the binary star Sirius. Properly called Sirius B, it is a tiny white dwarf star having the Sun's mass, but only about one-fiftieth of the Sun's diameter. It was the first white dwarf to be recognized.

communications blackout

A breakdown in communications that occurs during re-entry of a spacecraft into the Earth's atmosphere. It occurs, for about 20 minutes, because radio waves are unable to penetrate the layer of ionized air that builds up round the spacecraft.

communications satellite
Also called comsat; a satellite that relays radio and television programmes, telephone, Telex messages and many other kinds of coded data. Most international communications satellites are now located in geostationary orbit over the equator, above the Atlantic, Pacific and Indian Oceans. From these vantage points they can cover virtually the whole of the populated world. See also **Intelsat; Molniya.**

comparator
See **blink microscope.**

Compass See **Pyxis.**

Compasses See **Circinus.**

Complex 39 The main launch site at the Kennedy Space Center, scene of all the launchings of Apollo and now of the space shuttle. There are two launch pads, A and B, which are located over 3 miles (5 km) from the dominant feature of the complex, the Vehicle Assembly Building.

comsat
See **communications satellite.**

conjunction A situation when two or more bodies are aligned in space. Mercury and Venus are in inferior conjunction when they lie directly between Earth and the Sun, and are in superior conjunction when they lie on the opposite side of the Sun. The outer planets are in conjunction when they lie on the opposite side of the Sun. See also **opposition.**

constellation A pattern made by a group of bright stars. The stars are seldom, in fact, physically grouped together. They just happen to lie in the same direction in space. The ancients named the constellations after creatures and objects they thought the star patterns resembled. Today astronomers use the constellations as convenient reference for locating celestial bodies. Some 88 constellations are generally recognized.

continuous spectrum The spectrum produced when the light emitted by a body heated to incandescence is passed through a spectroscope. It shows an unbroken band of colour from red to violet. The interior of a star produces white light with such a spectrum. However, when the light passes through the star's outer atmosphere, certain wavelengths are absorbed, giving rise to a dark-line, or **absorption spectrum.**

Coon Butte crater An alternative name for the **Arizona meteor crater.**

co-orbitals Two tiny moons of

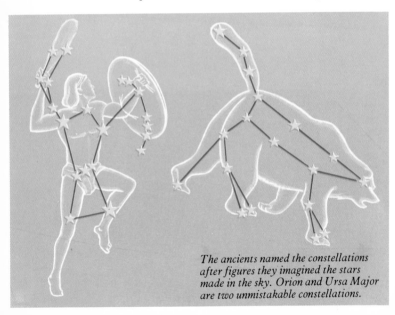

The ancients named the constellations after figures they imagined the stars made in the sky. Orion and Ursa Major are two unmistakable constellations.

Constellations of the northern hemisphere, from an early book on astronomy.

Saturn, discovered by the Voyager 1 space probe in 1980. Respectively about 180 and 120 km (110 and 75 miles) across, they orbit the planet together and could be two halves of a single moon that split in two.

coordinates
See **astronomical coordinates.**

Copernican system   Sun-centred view of the universe, as was put forward by Copernicus in 1543. Contrast **Ptolemaic system.**

Copernicus   One of the most conspicuous lunar craters, located to the north of Mare Imbrium. Some 90 km (56 miles) across, Copernicus has a central mountain range, and its terraced walls rise over 3900 m (13,000 ft) from the crater floor.

Copernicus (Spacecraft) The name of the third US orbiting astronomical observatory (OAO-3), which observed in the ultraviolet region between 1972 and 1981.

Copernicus, Nicolaus (1473–1543) Polish priest-astronomer who developed the Sun-centred view of the universe, against the accepted teachings of the Church. His views were expressed in a book published in the year he died entitled *De Revolutionibus Orbium Coelestium (Concerning the Revolutions of the Celestial Bodies).*

39

corona The outer layer of the Sun's atmosphere, visible as a pearly white halo usually only during a total solar eclipse. It extends many millions of kilometres out from the Sun.

Corona Australis The Southern Crown; a curve of faint stars forms this faint southern constellation, near Sagittarius.

Corona Borealis The Northern Crown; a small but easily located northern constellation close to Boötes. Its star T is an interesting example of a recurrent nova.

coronograph An instrument that, when used in a telescope, causes an artificial eclipse and permits study of the solar corona; invented by French astronomer Bernard Lyot in 1930.

Corvus The Crow; a small southern constellation which appears to be perched on Hydra.

COS-B A European astronomical satellite that observed at gamma-ray wavelengths between 1975 and 1982.

cosmic rays High-energy particles that bombard the Earth from outer space, particularly from the Sun as the solar wind. They consist mainly of protons and electrons. When they encounter the air in the upper atmosphere, other particles are created, such as mesons.

cosmogony The study of the origin and evolution of the universe.

cosmology The study of the universe as a whole.

cosmonaut The Russian name for an astronaut.

Cosmos A series of Russian scientific and research satellites. By 1985 over 1600 Cosmos satellites had been launched.

cosmos An alternative name for the universe.

coudé focus A method of focusing used in reflecting telescopes. Light is gathered by the main mirror, reflected back up the tube to a secondary mirror, then back down to a third mirror, which reflects the light

The Crab nebula in Taurus, the remains of a supernova in 1054.

Craters on Mercury, caused by the impact of meteorites aeons ago.

Robert Crippen (left) pictured in the cockpit of the first operational orbiter Columbia, with veteran astronaut John Young.

down through the telescope's polar axis. It is much used for spectroscopic work.

countdown The counting down of a certain period of time to the time of lift-off (T) of a launch rocket. 'T—2' would mean two hours before lift-off.

counterglow An alternative term for **gegenschein**.

Crab See **Cancer**.

Crab nebula (M1) An expanding nebula located in the constellation Cancer (the Crab). It is the remains of a spectacular supernova witnessed by Chinese astronomers in AD 1054. Up to 10 light-years across, it lies about 5000 light-years away. It contains an energetic pulsar.

Crab pulsar A neutron star in the Crab nebula, which has one of the fastest rotation rates known. Only some 20 km (12 miles) across yet more massive than the Sun, the Crab pulsar flashes on and off 30 times a second at X-ray, radio and optical wavelengths.

Crane See **Grus**.

Crater The Cup; a small, faint southern constellation containing little of interest.

crater A circular impression in the surface of a planet or moon. Most are impact craters, made by the bombardment of rocks from outer space. Others are volcanic in origin. Craters abound on the Moon, Mercury, Mars and the rocky satellites of the planets. The best preserved crater on Earth is the Arizona meteor crater.

crater rays Bright rays that fan out from large lunar craters, such as Tycho and Copernicus. Stretching for hundreds of kilometres, they are probably composed of shiny glassy material flung out when the crater was formed.

crêpe ring The transparent inner (C) ring of Saturn observed from Earth, some 15,000 km (12,000 miles) wide. Close-up photographs of the rings taken by Voyager show that an even fainter ring extends from the crêpe ring probably to the planet's cloud tops.

Crippen, Robert (born 1937) US pilot-astronaut who first flew into space (with John Young) on the maiden flight of the space shuttle Columbia on 12 April 1981.

Crow See **Corvus**.

Crux The Southern Cross; best known of the southern constellations, whose brilliant stars Alpha (or Acrux) and Gamma serve as pointers to the southern celestial pole. Two other interesting features of Crux are the Jewel Box and the Coal Sack.

cryogenic propellants Any rocket propellants that have to be stored at very low temperatures, such as liquid hydrogen and liquid oxygen (–253°C and –183°C respectively).

culmination The moment when a heavenly body crosses the observer's meridian. A circumpolar star will be observed in upper (superior) culmination when it transits between the celestial pole and the zenith, and in lower (inferior) culmination when it transits between the pole and the horizon.

Cup See **Crater.**

61 Cygni A double star in the constellation Cygnus, of historical interest in that it was the first star to have its parallax and distance directly measured. See **Bessel.**

Cygnids A meteor shower that occurs between about 18–22 August every year.

Cygnus The Swan; also called the Northern Cross; one of the most beautiful and easily recognizable northern constellations, which does look rather like the figure it is supposed to represent. It lies on the Milky Way. Its brightest star, the 1st mag Deneb, forms one corner of the Summer Triangle.

Cygnus A The second most powerful radio source in the heavens (after Cassiopeia A) in the constellation Cygnus. It is an enormous galaxy, 1000 times more massive than our own.

Cygnus X-1 A strong emitter of X-rays in the constellation Cygnus. Astronomers think that it is a binary star system, one component of which could be a black hole 10 times more massive than the Sun.

D

Daedalus A study project by the British Interplanetary Society in the 1970s to examine the feasibility of interstellar travel by a robot probe. A design was put forward that would be powered by nuclear pulse rockets, employing a series of mini H-bomb explosions.

dark nebula A dense cloud of gas and dust that blots out the light of any stars behind it. Two well known dark nebulae are the Horsehead nebula and the Coal Sack.

Darmstadt Site of the European Space Agency's centre for operating and communicating with spacecraft in orbit. It is known as ESOC, the European Space Operations Centre.

day The time it takes the Earth to spin once on its axis. The solar day is the Earth's period of rotation in relation to the Sun. For general time purposes we use a mean solar day of 24 hours. This represents the average solar day throughout the year. The actual solar day varies because the Earth's orbit is not circular, but elliptical. The true period of rotation of the Earth, in relation to the stars, is the sidereal day, 23 hr 56 min 4.1 sec. This is the time astronomers use.

DBS The abbreviation for **direct broadcasting satellite.**

Death Star See **Nemesis.**

declination One of the coordinates in the equator system of astronomical coordinates. It is the angular distance to a star measured from the celestial equator. It is equivalent to celestial latitude. Declination north of the equator is deemed positive; declination south of the equator negative.

The Daedalus interstellar probe, being assembled in space. It would use hydrogen fuel extracted from the atmosphere of Jupiter (top right).

deep space The space beyond the Earth's sphere of influence.

Deep Space Network A network of NASA tracking stations for communicating with distant space probes. The network includes 64-m, 34-m and 26-m (200 ft, 110-ft and 85-ft) dish antennae at Goldstone in California, Canberra in Australia and Madrid in Spain.

deferent and epicycle A system for describing planetary motions around the Earth used by ancient astronomers. It tried to account for the peculiar movement of the planets, as seen from the Earth. In the system a planet travels in a circle (epicycle), whose mid-point itself travels in a circle (deferent) around the Earth.

Deimos The smaller and more distant of the two moons of Mars. An irregular shaped pitted lump of rock about 12 km (7 miles) across, it orbits at a distance of some 23,500 km (14,500 miles).

Delphinus The Dolphin; a small constellation on the edge of the Milky Way close to Altair.

Delta One of the most successful US launching rockets, operating since 1962. The present version is some 35 m (116 ft) long and consists of a three-stage core and nine 'strap-on' solid rocket boosters.

Delta Cephei A variable star in the constellation Cepheus whose brightness changes precisely over a precise period of time. It is the prototype of the **Cepheids.**

Deneb (Alpha Cygni) A very hot supergiant star (A2), the brightest in the constellation Cygnus and at mag 1.25 the 19th brightest in the heavens. It lies much farther away (1630 light-years) than all the other very bright stars and has an absolute magnitude of –7.3.

Diamant A rocket that launched France's first satellite in November 1965, making it the world's third space power.

diamond-ring An aptly named effect seen at the beginning or end of a total solar eclipse. The 'diamond' is the last flash of sunlight before totality or the first flash after totality.

dichotomy The phase of the Moon, Mercury or Venus, when exactly half the surface is lit, and the terminator is a straight line.

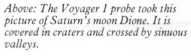

Below: The tracking station at Goldstone in California, part of NASA's Deep Space Network for tracking interplanetary probes.

Above: The Voyager 1 probe took this picture of Saturn's moon Dione. It is covered in craters and crossed by sinuous valleys.

diffraction grating A device used in spectroscopes to split up light into a spectrum. A reflection grating consists of a sheet of metal on which fine parallel lines have been ruled, 6000 or more to the centimetre (15,000 or more to the inch). Transmission gratings have lines ruled on sheets of glass.

Dione This is one of Saturn's larger moons, measuring about 1120 km (700 miles) across. Composed of rock and ice, it has a cratered surface.

direct broadcasting satellite Or DBS; a communications satellite powerful enough to broadcast TV programmes directly into the home via small dish antennae. It is located in geostationary orbit.

The shuttle orbiter Discovery stands on the launch pad at the Kennedy Space Center prior to its maiden voyage in 1984. It is mounted on its huge external fuel tank and rocket boosters.

direct motion Motion of a heavenly body in an anticlockwise direction when observed from celestial North, that is from West to East. All the planets have direct motion around the Sun. A few planetary moons have an opposite, or retrograde motion.

Discovery The third operational space shuttle orbiter, which first blasted off into space in August 1984, on mission 41-D. Its crew included the second US woman astronaut Judy Resnik.

dispersion The splitting up of light into its component colours, which occurs when it passes through a prism or diffraction grating.

distances in space are so vast that they are meaningless when they are expressed in terrestrial units such as kilometres. One convenient unit astronomers use is the light-year, the distance light travels in a year. The nearest star then becomes just over 4 light-years away; the farthest quasars over 15,000 million light-years (which would be some 1.5×10^{23} kilometres!). An alternative unit is the parsec. Within the solar system, the astronomical unit is often used.

diurnal motion The daily rotation of the heavens around the celestial poles from East to West. It is an apparent motion caused by the spinning of the Earth on its axis from West to East.

docking The linking up of two spacecraft in space.

Dog Star A common name for brightest star in the sky, Sirius, in the constellation Canis Major, the Great Dog.

Dollond, John (1706–1761) English optician who first produced achromatic lenses for telescopes, which could cure the colour blurring of chromatic aberration.

Dolphin See **Delphinus.**

Donati, Giovanni (1826–1873) Italian astronomer who discovered several comets, including the very bright comet of 1858, which now bears his name.

Doppler, Christian (1803–1853) Austrian physicist who in 1842 first described the Doppler effect, so important in the study of the spectral lines of starlight.

Doppler effect The apparent change in the frequency (or wavelength) of sound or light when the object emitting the sound or light is moving relative to the observer. With a star, the Doppler effect causes the spectral lines in the star's spectrum to shift position – towards the blue end of the spectrum if the star is approaching us, or towards the red if the star is receding. See **blue shift; red shift.**

Dorado The Swordfish; not a particularly interesting southern constellation, except that it contains part of the Large Magellanic Cloud.

double star A pair of stars that appear close together in the sky. Optical doubles are stars that just happen to lie together in our line of sight. Double stars that are physically associated are called binaries.

Dove See **Columba.**

downlink Communications broadcast from a spacecraft to the ground.

Draco The Dragon; a long winding northern constellation, not easy to make out. Interestingly, Alpha Draconis, or Thuban, was the Earth's pole star when the Great Pyramid of Giza was constructed.

drag The resistance experienced by a body moving through the air. Spacecraft re-entering the atmosphere from orbit use aerodynamic drag to slow them down (see **re-entry**).

Dragon See **Draco.**

Draper classification Also called Harvard Classification; a method of classifying stellar spectra developed by Annie Jump Cannon at the Harvard College Observatory. The main spectral classes are O,B,A,F,G, K,M,R and N, which can be remembered in order by the unforgettable mnemonic: 'Oh, Be A Fine Girl, Kiss Me Right Now!'

Dreyer, Johan Ludwig Emil (1852–1926) Danish astronomer who became director of the Armagh Observatory, Ireland. He developed the New General Catalogue (NGC) system of classifying nebulae and star clusters, which is still used today.

Dryden Flight Research Facility A NASA establishment for high-speed flight testing located in the Mojave Desert, some 60 miles (100 km) north of Los Angeles in California. It is adjacent to the Edwards Air Force Base, whose runway is often used for space shuttle landings.

DSN Abbreviation for NASA's **Deep Space Network.**

Dumb-bell nebula (M27) An attractive planetary nebula in the constellation Vulpecula. About one light-year across, it lies about 700 light-years away.

dwarf galaxy A galaxy that is dimmer and smaller than usual. It is typically a few thousand light-years across and generally elliptical or irregular.

dwarf nova A kind of variable star that shows spectacular changes in

Apollo astronauts took this superb picture of the full Earth.

This Landsat picture of Earth shows evidence of human habitation.

brightness rather like a nova. The star U Geminorum is typical and representative of a class of some 200 stars. It brightens every 100 days or so by 5 magnitudes or more.

dwarf star A star of relatively small dimensions. The Sun is classed as a yellow dwarf. Red dwarfs are much smaller and of low brightness. White dwarfs are smaller still and of enormous density.

Dyson sphere A concept of planetary engineering advanced by Freeman Dyson at Princeton University's Institute for Advanced Study. It is a ring of matter from dismantled planets moved into orbit around a parent star. It would be one way of trapping the maximum amount of energy from the star.

E

E=mc^2 The mass-energy equation advanced by Albert Einstein, which demonstrates the equivalence of mass and energy. It shows what enormous energy (E) is released when, as happens in nuclear reactions, a certain amount of mass (m) is destroyed; c is the velocity of light.

Eagle See **Aquila**.

Eagle (Spacecraft) The code name for the Apollo 11 lunar module that took astronauts Armstrong and Aldrin down to the Moon on 20 July 1969.

Early Bird The first communications satellite launched by Intelsat, which went into service in geostationary orbit in June 1965 with 240 telephone circuits, linking Europe and North America.

Earth Our home planet, the only one in the solar system to provide the conditions suitable for life as we know it. The third planet out from the Sun, it lies on average 150 million km (93 million miles) away. The diameter at the equator (12,756 km, 7,926 miles) is 43 km (26 miles) greater than that at the poles. The thin solid crust, up to 30 km (20 miles) thick, 'floats' on a semi-molten mantle and is split up into a number of 'plates'. These plates

49

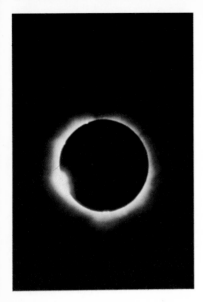

Above: A total eclipse of the Sun. The Moon exactly covers the Sun's disc for a few minutes.

Below: In an eclipsing binary, one star passes in front of the other from time to time.

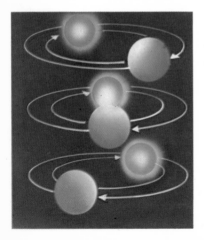

are continually on the move, accounting for the phenomenon of continental drift. The inner core of the Earth is thought to be made up of iron and nickel, which gives rise to the Earth's magnetic field.

earth-grazer An asteroid, such as Hermes, that passes relatively close to the Earth.

Earth Resources Technology Satellite (ERTS) See **Landsat.**

Earth rotation synthesis An alternative name for **aperture synthesis.**

earthshine Also called ashen light; a faint illumination of the dark part of the Moon near New Moon phase, caused by the reflection of light from the Earth.

earth station A ground station for relaying signals to and from a communications satellite. Britain's main earth stations are at Goonhilly in Cornwall and Madley, near Hereford.

Eastern Test Range A term often used for the Cape Canaveral rocket launching site.

eccentricity A measure of how elongated an elliptical orbit is. Among the planets, Venus (0.0068) has the least eccentric orbit, that is, the most nearly circular, while Pluto has the most eccentric orbit (0.25).

Echo A series of aluminium-coated balloons that were used as the first communications satellites. They were passive, merely bouncing signals between earth stations. Echo 1 (1960) measured some 30 m (100 ft) across.

eclipse The passage of one heavenly body in front of another so as to obscure it. An eclipse of the Sun occurs when the Moon passes in front of it (see **solar eclipse**). An eclipse of the Moon occurs when the Moon passes into the Earth's shadow in space (see **lunar eclipse**).

eclipsing binary A kind of binary star in which one component periodically passes in front of the other in our line of sight. This produces a periodic dip in brightness, giving the system the appearance of a variable star. See **Algol.**

ecliptic The apparent path of the Sun across the celestial sphere during the year. It is inclined $23\frac{1}{2}°$ to the celestial equator, which it intersects at the First point of Aries and the First point of Libra.

ECS A series of European communications satellites, the first of which was launched by Ariane in 1983.

Edwards Air Force Base
A US base north-east of Los Angeles which is an alternative landing for the space shuttle.

Effelsberg radio telescope The largest fully steerable radio telescope, with a dish 100 m (330 ft) across. It is located near Bonn, West Germany.

Einstein, Albert (1879–1955)
German physicist who revolutionized science with his theories of relativity, dating from 1905, and his concepts of space and time. To the general public he is best known for his mass-energy equation $E = mc^2$, in which he demonstrated that mass and energy are equivalent. This helps explain the energy release in nuclear reactions.

The Effelsberg radio telescope, the largest fully steerable dish in the world.

Above: This picture, taken by the Einstein Observatory, shows the Andromeda galaxy. The bright points are sources of strong X-radiation.

Below: An elliptical galaxy, which lacks the arms of a spiral galaxy.
Right: Saturn's Moon Enceladus, pictured by the Voyager 2 probe.

Einstein Observatory The name given to the second US High Energy Astronomical Observatory (HEAO–2) launched in 1979. Weighing some 3 tonnes and nearly 6 m (20 ft) long, Einstein scanned the heavens for $2\frac{1}{2}$ years at X-ray wavelengths.

Elara One of Jupiter's tiny moons, little more than 80 km (50 miles) across. It orbits some 11,700,000 km (7,300,000 miles) from the planet.

ELDO Abbreviation for European Launcher Development Organization, a body that became integrated into the present European Space Agency.

electromagnetic spectrum
The whole range of electromagnetic radiations that bodies such as stars give out. The radiations differ primarily in wavelengths (or frequency). Going from short to long wavelengths, the main radiations are gamma-rays, X-rays, ultraviolet rays,

visible light, infrared rays, microwaves and short, medium and long radio waves.

electron One of the fundamental atomic particles, present in all atoms in a so-called shell surrounding the atomic nucleus. It has a negative electron charge and has a mass less than 1/1800th that of a proton.

elliptical galaxy A major class of galaxy with the shape of an ellipse. Such galaxies vary from near spheres to extremely elongated shapes, categorized from EO to E7.

elliptical orbit An orbit in the shape of an ellipse. The planets have elliptical orbits around the Sun, as do moons around their planets and satellites around the Earth. Kepler was the first to realise that orbits in space were elliptical not circular.

elongation The difference in celestial longitude between the Sun and a planet. It is often expressed as the planet's angular distance from the Sun.

emission nebula A bright nebula that emits light of its own as its gases are excited by radiation from nearby stars.

emission spectrum The spectrum produced by an incandescent element which is unique to that element. It consists of bright coloured lines against a dark background.

Enceladus Saturn's second closest moon, some 500 km (300 miles) across, which orbits about 240,000 km (150,000 miles) away. Highly reflective because of ice, its surface is very young.

Encke's comet A regular comet that has the shortest period, or return time – 3.3 years.

Encke's division An apparent gap in Saturn's ring system, near the outer edge of the A ring. As with the wider Cassini division, it still contains a number of ringlets.

encounter The close approach of a space probe to its target body, usually a planet.

The prototype space shuttle orbiter Enterprise swoops down towards the runway during landing tests.

Enterprise Name of the prototype space shuttle orbiter. It flew in atmospheric approach and landing tests in 1977, but will never fly in space. It was named after the spaceship in the 'Star Trek' TV series.

ephemeris An astronomical calendar listing the positions of the heavenly bodies throughout the year and other relevant information. Most governments publish ephemerides (the plural) for such purposes as navigation. In Britain the publication is *The Astronomical Ephemeris.*

epicycle
See **deferent and epicycle.**

epoch A date used as a reference point in star catalogues, for example to quote stellar coordinates, which change gradually over the years. The latest catalogues use epoch 2000.

equation of time The difference between mean (i.e. clock) time and solar (i.e. sundial) time.

equatorial mounting The most common method of mounting a telescope: on a polar axis that points to the celestial pole, and a declination axis at right-angles to it. The telescope is turned about the polar axis to follow a star across the sky.

equinoxes The two times of the year when the lengths of night and day are equal. Astronomically this occurs when the Sun is on the celestial equator, that is, when the plane of the ecliptic intersects the celestial equator. This happens on about March 21 (vernal or spring equinox) and September 23 (autumnal equinox).

Equuleus The Colt; a tiny and inconspicuous constellation between Pegasus and Delphinus.

Eratosthenes (275–192 BC) Greek astronomer who measured the size of the Earth with remarkable accuracy.

The main satellite control centre of the European Space Agency, at Darmstadt in West Germany.

Eridanus A long winding constellation that extends from the mag 1 Achernar in the far south up to near Rigel in Orion on the celestial equator.

Eros An asteroid that comes relatively close to Earth (about 25 million km, 16 million miles). It is thought to be an elongated lump of rock no more than 35 km (21 miles) long.

EROS Data Center US distributor of remote-sensing data and pictures from satellites and aircraft. It is the world's chief distributor of Landsat data. It is located at Sioux Falls in South Dakota.

ERTS Earth Resources Technology Satellite. See **Landsat.**

ESA The European Space Agency, an organization that coordinates space projects in Europe. It has 11 members – Belgium, Britain, Denmark, France, West Germany, Ireland, Italy, the Netherlands, Spain, Sweden and Switzerland. See also **Ariane.**

escape velocity The speed a spacecraft requires to escape from a body's gravity. The Earth's escape velocity is 40,000 km/h (25,000 mph). That of the Moon is only 8500 km/h (5300 mph), while that of Jupiter is 220,000 km/h (140,000 mph).

ESOC See **Darmstadt.**

ESRIN An ESA centre at Frascati in Italy that collects and distributes satellite data.

ESRO The European Space Research Organization, a forerunner of **ESA.**

ESSA The US Environmental Science Services Administration's series of weather satellites.

ESTEC An ESA research and test centre at Noordwijk in the Netherlands; the European Space Research and Technology Centre.

ether See **aether.**

Eudoxus of Cnidus (*ca* 400–350 BC) Greek mathematician and astronomer, who tried to explain the observed motions of the Moon and planets quantitatively for the first time, using a system of 27 spheres.

Europa Jupiter's second largest satellite, measuring 3126 km (1942 miles) across, which orbits some 670,000 km (420,000 miles) from the planet. Its smooth ice-covered surface is crossed with fracture lines.

Eutelsat The European Telecommunications Satellite Organization.

EVA This abbreviation stands for **extravehicular activity.**

evening star The name given to the planet Venus when it becomes prominent in the West at sunset.

event horizon The boundary of a black hole.

exobiology The study of the possibilities of life elsewhere in the universe, outside the Earth.

Exosat A European X-ray satellite launched in 1983.

exosphere The outer part of the Earth's atmosphere, which gradually merges into space. It is generally considered to extend from about 500 km (300 miles) up.

expanding universe A generally accepted concept that the universe is expanding in size. This is observed in the outrush of the galaxies, which are almost all travelling away from us, the farthest ones being the fastest. This expansion is thought to have started as a result of the 'Big Bang', which brought the universe into being some 15,000 million years ago. See **Big-Bang theory.**

exploding stars See **neutron star; nova; supernova.**

Explorer 1 The first US satellite, which was launched by a Juno rocket from Cape Canaveral on 31 January 1958. Orbiting at heights up to 2500 km (1500 miles), Explorer 1 discovered the Van Allen radiation

Left: A man in Japan reported seeing this extraterrestrial being in 1978, and claimed it passed him telepathic messages.

Above: Apollo 12 astronaut taking part in extravehicular activity on the Moon in November 1969, during the second Moon landing.

belts. It remained in orbit until April 1970 after travelling 58,376 times around the Earth.

extragalactic nebulae
An early term for **galaxies.**

extraterrestrial Existing outside the Earth. Extraterrestrial intelligence refers to beings from other worlds.

extravehicular activity (EVA) Activity by astronauts outside a spacecraft. The correct term for what is popularly called spacewalking.

extravehicular mobility unit (EMU) The term NASA uses for the shuttle crew's spacesuits.

eyepiece Also called an ocular. The lens in a telescope that is closest to the eye. Its function is to magnify the image formed by the telescope's objective lens. It usually consists of a pair of lenses (eyelens and field lens).

F

---

faculae Bright areas of the Sun's surface, usually occurring in the vicinity of sunspots and often before they appear.

falling star
A common name for a **meteor.**

false-colour imagery A computerized technique for producing pictures in artificial colours so as to make certain features stand out.

finder A low-powered telescope with a wide field of view, mounted on a larger one to help locate quickly the object for observation.

fireball See **bolide.**

fireball radiation An alternative name for **background radiation.**

First point of Aries One of the two points on the celestial sphere where the plane of the ecliptic crosses the plane of the celestial equator. The Sun is located at this point at the vernal equinox (about March 21). The first point of Aries is the zero mark for the measurement of right ascension, or celestial longitude.

First point of Libra One of the two points on the celestial sphere where the plane of the ecliptic crosses the celestial equator. The Sun is located at this point at the autumnal equinox (about September 23).

First Quarter The phase of the Moon after New Moon, when the surface is half illuminated.

Fishes See **Pisces.**

fixed star A term the ancients used to distinguish proper stars, 'fixed' to the celestial sphere, from the 'wandering stars', or planets.

Flamsteed, John (1646–1719)
English astronomer who in 1675 became the first Astronomer Royal, at the newly built Greenwich Observatory. He compiled an accurate star catalogue.

flare, solar A sudden burst of radiation from the Sun, usually in the region of sunspots, when radio waves, X-rays and charged particles are given off. On Earth, flares give rise to aurorae and magnetic storms.

flare stars Irregular variable stars, usually red dwarfs, that suddenly increase in brightness and radio emissions. UV Ceti is typical.

flash spectrum The fleeting bright-line emission spectrum of the chromosphere just before and just after totality during a solar eclipse.

flocculi Also called plages; small cells of hotter and cooler gas on the Sun, giving it a mottled appearance in monochromatic photographs.

Fly See **Musca.**

fly-by A planetary encounter in which a probe makes observations as it flies past a planet.

Flying Fish See **Volans.**

Flying Horse See **Pegasus.**

Left: A false-colour image of Comet Kohoutek, 1973.

flying saucer A UFO with a characteristic saucer shape.

flyswat A makeshift device with which astronauts on the 51–D shuttle mission (April 1985) tried to reactivate a 'dead' communications satellite.

Fomalhaut (Alpha Piscis Austrini) A brilliant star (A1) in the southern constellation Piscis Austrinus. At mag 1.1 it is the 18th brightest star in the sky.

Fornax The Furnace; a southern constellation showing little of interest.

Foucault pendulum A long pendulum used to demonstrate that the Earth rotates on its axis. The direction of swing appears to change gradually, but it is actually the Earth that is moving beneath the pendulum. The French physicist Jean Foucault first demonstrated the effect in 1851.

Fox See **Vulpecula.**

Fraunhofer, Joseph von
(1787–1826) German astronomer who investigated and explained the dark lines in the Sun's spectrum, first observed by W.H. Wollaston in England in 1802.

Below: This 'flying saucer' was supposedly photographed in Peru.

Fraunhofer lines Dark lines in the spectrum of sunlight first explained by Joseph von Fraunhofer. They are caused by the absorption of certain wavelengths by gases in the Sun's cooler outer atmosphere.

free fall A state in space when a body is moving freely under gravity. Spacecraft in orbit are in free fall, as is everything inside them. Because of this everything appears to have no weight, leading to the popular notion of weightlessness, or zero-G.

Friendship 7 The Mercury capsule in which astronaut John Glenn became the first American to orbit the Earth, on 20 February 1962.

F stars Yellowish-white stars of spectral type F, with a surface temperature of about 7000°C.

fuel cell A cell that produces electricity by catalytically combining hydrogen and oxygen gases to form water. The space shuttle uses batteries of fuel cells to produce power and provide drinking water for the crew.

Full Moon The phase of the Moon, at age $14\frac{1}{2}$ days, when the surface facing the Earth is completely illuminated. It is not the best time for observing the Moon, though crater rays show up best at this time.

Furnace See **Fornax.**

fusion See **nuclear energy.**

G

g or **G** The acceleration due to gravity, which acts on all matter on or near the Earth to produce the sensation of weight. It is 981 cm (32.2 ft) per second per second.

Gagarin Cosmonaut Training Centre The chief training centre for Russian cosmonauts, at Zvezdnyy Gorodok, or Star City, near Moscow.

Gagarin, Yuri Alekseyevich (1934–1968) Russian cosmonaut who pioneered manned spaceflight on 12 April 1961. He made one orbit of the Earth in his capsule Vostok 1.

galaxies Systems of stars, which typically contain hundreds of thousands of millions of members. There are three main types of galaxies, classified according to their appearance – elliptical galaxies, spiral galaxies and barred spiral galaxies. The stars within galaxies often gather together in open clusters and globular clusters. Clouds of dust and gas can often be found between the stars (see **nebula**).

Galileo Galilei (1564–1642)
Italian astronomer who in 1609 made the first telescopic observations of the heavens. His observations of Jupiter's satellites and the phases of Venus convinced him of the Copernican view of a solar system.

Left: Cosmonaut Yuri Gagarin, pictured in his spacesuit before his pioneering orbital flight in 1961.

Opposite: John Glenn climbs into his Mercury capsule before his first flight.

Below: A typical spiral galaxy in Canes Venatici.

Galaxy, The The galaxy to which our Sun belongs, also called the Milky Way galaxy. It is a typical spiral galaxy of the Sb type, and contains something like 100,000 million stars. It measures about 100,000 light-years across, and the Sun is located on a spiral arm some 30,000 light-years from the centre. The central bulge measures about 20,000 light-years across.

Galilean satellites The four large satellites or moons of the planet Jupiter, discovered by Galileo in 1610. They are, in order going out from the planet, Io, Europa, Ganymede (the largest) and Callisto.

Galilean telescope The type Galileo made, which has a convex objective and a concave eyepiece. This produces an erect (right-way-up) image. The principle is still used in opera glasses but not in astronomical telescopes.

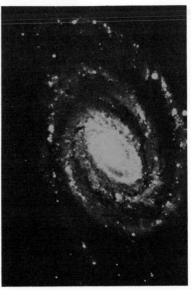

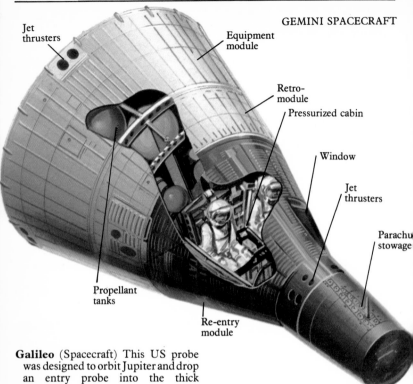

GEMINI SPACECRAFT

Jet thrusters

Equipment module

Retro-module

Pressurized cabin

Window

Jet thrusters

Parachu stowage

Propellant tanks

Re-entry module

Galileo (Spacecraft) This US probe was designed to orbit Jupiter and drop an entry probe into the thick atmosphere.

Galle, Johann Gottfried (1812–1910) German astronomer who discovered an eighth planet, Neptune, in 1846.

gamma-ray astronomy Studying the short wavelength gamma-rays that heavenly bodies give out. This must be done by satellites, since gamma-rays are abosrbed by the atmosphere. Two of the most brilliant gamma-ray sources found so far are the Crab and Vela pulsars.

Gamow, George (1904–1968) American astronomer who championed and developed Lemaître's idea of the evolutionary, Big-Bang origin of the universe.

gantry Another name for the access tower on a rocket launch pad.

Ganymede The largest satellite of Jupiter and of the whole solar system. Measuring 5276 km (3278 miles) across, it is made up of rock and ice. It orbits about 1 million km (600,000 miles) out.

gegenschein Also called counter-glow; a faint glow visible in the night sky immediately opposite to the direction of the sun. It is related to the zodiacal light.

Gemini The Twins; a splendid northern constellation of the zodiac dominated by the bright 'twin' stars Castor and Pollux.

Geminids A meteor shower with the radiant in Gemini, often rich in fireballs, occurring during the second week in December.

Gemini project A series of 10 US missions in two-man craft, beginning with Gemini 3 in March 1965. On Gemini 4, three months later, Edward White made the first US spacewalk. The final mission, Gemini 12, was in November 1966.

geocentric Centred on the Earth.

geostationary orbit Also called Clarke orbit; an orbit 35,900 km (22,300 miles) high above the Earth in which the orbital period is exactly 24 hours. A satellite placed in this orbit over the equator and travelling in the direction of the Earth's rotation appears to be stationary, because it turns at the same rate as the Earth.

getaway special A small payload taken up in the space shuttle in a canister in the payload bay. Getaway specials enable universities and other research bodies to fly small instrument packages in space at modest cost.

G-forces The forces of acceleration and retardation experienced by astronauts during lift-off and re-entry. Shuttle astronauts 'pull' a little more than 3Gs (three times the normal force of gravity) during lift-off.

giant star A bright star typically tens of times bigger than the Sun and of low density. See **red giant.**

gibbous The shape of a body when more than half but not all of its surface is illuminated. The Moon is gibbous between quarter and full phases.

Giotto probe A spacecraft launched by ESA in 1985 to rendezvous with Halley's comet on its 1986 return.

Giraffe See **Camelopardalis.**

Left: The two-man Gemini capsule, in which American astronauts travelled in space in 1965/66.

Below: The Giotto probe heads for its 1986 rendezvous with Halley's comet.

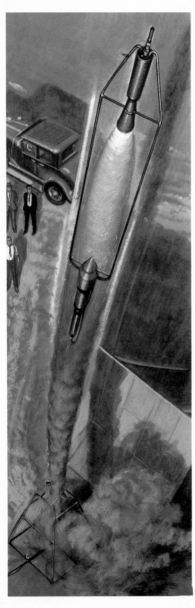

Glenn, John H. (born 1921) US astronaut who became the first American to orbit the Earth on 20 February 1962 in the Mercury space capsule Friendship 7. He made three orbits.

global positioning system A US satellite navigation network consisting of 18 Navstar satellites. They carry superaccurate atomic clocks and emit signals from which ships can pinpoint their position to within 15 m (50 ft).

globular cluster A globe-shaped cluster of tens of thousands of stars usually found around the centre of a galaxy. They are mainly old (population II) stars. Omega Centauri and 47 Tucanae in the southern hemisphere, and M13 in Hercules in the northern hemisphere are magnificent globular clusters visible to the naked eye.

GMS This Japanese geostationary meteorological satellite is of similar design to GOES.

gnomon At its simplest, a vertical stick that casts a shadow. The sundial has a gnomon. The Apollo astronauts also used a gnomon to indicate Sun angles on the Moon.

Goddard, Robert H. (1812–1945) US rocket pioneer who launched the world's first liquid propellant rocket on 16 March 1926. It used petrol and liquid oxygen as propellants.

Goddard Space Flight Center NASA facility located at Greenbelt, Maryland, about 16 km (10 miles) from Washington DC. It is NASA's main communications centre, and is responsible for collecting and distributing satellite data.

GOES Geostationary operational environmental satellite; a series of US meteorological satellites which are located in geostationary orbit over the Atlantic, Pacific and Indian Oceans. They are operated by NOAA.

Goldstone An earth station in Gold-

Opposite: Robert Goddard's first liquid-propellant rocket soars up to 60 m (200 ft) in the air.

Above: The magnificent globular cluster M13, in the constellation Hercules.

stone, California, which is the main US link in NASA's Deep Space Network.

Gold, Thomas (born 1920) US astronomer who is one of the authors (with Hoyle and Bondi) of the steady-state theory of the universe.

Goodricke, John (1764–1786) English astronomer best known for his work on variable stars. He first explained the variation of Algol as an eclipsing binary, in 1782.

granulation The grain-like structure of the Sun's visible surface, the photosphere. It is caused by pockets of gas continually rising and falling.

gravitation The force of attraction that exists between any bodies with mass. It is one of the fundamental forces of the universe. Isaac Newton put forward his theory of gravitation in 1687. Mathematically, the force (F) between two bodies of mass m_1 and m_2 a distance d apart is given by $F = Gm_1m_2/d^2$, where G is the gravitational constant.

gravitational lens An effect caused by the gravity of a massive body bending light which passes it. When this occurs, double images of the objects emitting the light may be observed. This accounts for the twin images of the so-called double quasar 0957+561.

graviton The hypothetical quantum of gravitational energy, analogous to the photon of light energy.

gravity The gravitational attraction of the Earth. The downward force a body on Earth of mass m experiences (its weight) equals mg where g is the acceleration due to gravity, which is 981 cm (32.2 ft) per second per second.

gravity assist A method employed in space flight that uses the gravity of one planet to accelerate a spacecraft towards another. The Voyager probes, for example, used the gravity of Jupiter to accelerate and redirect them to Saturn. This method greatly reduces journey times.

gravity waves Hypothetical waves responsible for the phenomenon of gravitation. Several people have built gravity wave detectors, but have been unable to detect them, if they exist.

Great Bear See **Ursa Major**.

great circle Any circle on the surface of a sphere (eg the celestial sphere), whose plane passes through the centre of the sphere.

Great Dog See **Canis Major**.

green flash A brief flash of green light sometimes observed just before the last vestige of the Sun's disc disappears at sunset. It is an atmospheric effect.

greenhouse effect An atmospheric condition, particularly noticeable on Venus, which has a heavy carbon dioxide atmosphere. The atmosphere lets in heat from the Sun but then traps it, like a greenhouse does. On Venus this effect is responsible for the very high temperature at the surface, over 450°C.

Greenwich Mean Time (GMT) The time along the Greenwich meridian, taken as the world time standard.

Greenwich meridian The prime meridian of the Earth, which passes through Greenwich, in London. It marks the zero point for measurement of longitude.

Greenwich Observatory
See **Royal Greenwich Observatory**.

Gregorian calendar The present calendar, named after Pope Gregory XIII who corrected the calendar existing in 1582 by omitting the 10 days between 4 October and 15 October. This wiped out an accumulated error in the Julian calendar. Errors since then have been avoided by deeming that century years are only leap years when divisible by 400 (eg 1600).

Gregorian telescope A reflecting telescope suggested by James Gregory (1638–1675) that predated Newton's reflector by five years. It was of similar design to the present Cassegrain telescope, except that the secondary mirror was concave not convex.

Grissom, Virgil I. (1926–1967) Pioneering US astronaut who made a Mercury suborbital flight in 1961 and a Gemini flight in 1965. Tragically he was burned to death (with Edward White and Roger Chaffee) in a fire inside an Apollo spacecraft during training.

Grus The Crane; quite a conspicuous southern constellation not far from Fomalhaut.

G stars Stars of spectral type G, which includes the Sun. They give off yellowish light and have a surface temperature of about 6000°C.

G-suit A pressure suit worn by pilots and astronauts that squeezes the lower part of the body and helps prevent the blood leaving the upper part under high G-forces.

Opposite: The planet Venus, which suffers severely from the greenhouse effect.
Below: Astronaut Virgil Grissom.

guest star
An old name for a supernova.

Guiana Space Centre The main launching site of the European Space Agency, at Kourou in French Guiana on the north-east coast of South America.

gyro An abbreviation for gyroscope. Gyros are widely used in the navigation and guidance systems of rockets and spacecraft.

H

Hadar (Beta Centauri) A brilliant white-hot star (B1) in the constellation Taurus. At mag 0.6 it is the tenth brightest star in the heavens. It lies some 390 light-years away.

Hale, George Ellery (1868–1938) US astronomer who invented the spectroheliograph for studying the Sun's spectrum. The 200-inch (5-m) telescope on Mt Palomar is named after him.

Hale Observatories The joint name for the Mount Wilson and Palomar observatories in southern California.

Hale telescope For many years the largest optical telescope in the world, located on Mt Palomar in California. Operational since 1948, it has a 200-inch (508-cm) diameter mirror. It was eclipsed in size in 1974 by the 6-m (236-inch) Zelenchukskaya reflector in the Caucasus.

Hall, Asaph (1829–1907)
US planetary astronomer perhaps best known for his discovery of the two tiny moons of Mars, Phobos and Deimos, in 1877.

Halley, Edmond (1656–1742)

English astronomer who became the second Astronomer Royal in 1720. He is best remembered for his prediction of the return of a comet he saw in 1682. It duly appeared in 1758, and henceforth became known as Halley's comet.

Halley's comet A famous periodic comet named after Edmond Halley, who first calculated its period (76 years) and successfully predicted its return in 1758. Records of the

ESA's Guiana Space Centre, launch site for the Ariane rocket.

Halley's comet, pictured at its 1910 return. It became optically visible again in 1985 as it neared the Sun on its 1986 return.

sightings go back to 240 BC, and it is depicted on the Bayeux Tapestry during its 1066 appearance. This century it appeared in 1910, and was recovered (detected) again in 1982 as it headed for its 1986 perihelion.

halo A luminous ring often seen around the Sun and the Moon. It is caused by the presence of ice crystals in the atmosphere, which scatter the light.

halo, galactic A sphere of mainly old (population II) stars that surrounds the centre of the Galaxy. It includes globular clusters and gas clouds.

Hare See **Lepus**.

Harvard classification
See **Draper classification**.

harvest moon The full moon closest to the autumnal equinox (September 23). At this time the full moon rises at much the same time on successive evenings.

HEAO US spacecraft, the High Energy Astronomical Observatory. Three HEAOs were launched in the 1970s, of which the most successful, known as HEAO-2 or Einstein, revolutionized X-ray study of the universe, scanning the heavens for two-and-a-half years from 1979 at X-ray wavelengths.

69

Left: The tile heat shield can be clearly seen on the shuttle orbiter Discovery.

Right: William Herschel was an outstanding astronomer, who built a 1.2 m (48-inch) reflector.

heat shield A coating on the outside of a spacecraft that protects it from the aerodynamic heating caused by friction during re-entry. The shuttle orbiter uses heat-resistant tiles as a heat shield.

helical rising The rising of a star at the same time as the Sun.

heliocentric Centred on the Sun.

heliostat A mirror in a solar telescope, that turns to reflect the Sun's light in a fixed direction.

helium After hydrogen, the most abundant element in the universe. In stars, hydrogen fuses into helium in nuclear reactions that provide the energy to keep the stars shining.

Hellas A large basin on Mars, measuring some 2000 km (1200 miles) across and 4 km (2.5 miles) deep, caused by the impact of a giant meteorite.

hemisphere Half of the Earth, or of the celestial sphere. The northern and southern hemispheres of the celestial sphere are divided by the celestial equator.

Herbig–Haro object A quite small nebula-like object that appears to be in an early stage of star formation.

Hercules A large but not very distinctive northern constellation. One of its most interesting features is the fine globular cluster M13, which is just visible to the naked eye.

Herdsman See **Boötes.**

Hermes A tiny asteroid, only about 1 km (0.6 miles) across, which strayed within 800,000 km (500,000 miles) of the Earth in 1937. Astronomically, this was close.

Herschel, William (1738–1822) German-born English astronomer who is often called the 'father of stellar astronomy'. He discovered Uranus in 1781. His only son John (1792–1871) was also a noted astronomer, whose catalogue of nebulae and star clusters formed the basis, via Dreyer, for the NGC classification.

Hertzsprung, Ejnar (1873–1967) Danish astronomer who discovered dwarf and giant stars. Independently of Henry Norris Russell, he devised a diagram relating stellar magnitude and spectral class.

Hertzsprung-Russell diagram A diagram devised independently by the Danish astronomer Ejnar Hertzsprung and the American astronomer Henry Norris Russell. It shows the relationship between the absolute brightness (magnitude) of a star and its spectral class, or

temperature. On the diagram, stars fall into distinct groups (see diagram). Most fall into a so-called main sequence, cutting the diagram diagonally.

Hesperus A name for the planet Venus when it is an evening star in the western sky.

Hewish, Anthony (born 1924) British astronomer who, working at Cambridge, discovered the first pulsar in 1967.

Hidalgo An asteroid (no. 944) that strays beyond the orbit of Saturn. Some 30 km (20 miles) across, it could be the remains of a comet.

Himalia One of the middle-distant group of Jupiter's moons, orbiting

Simplified version of the Hertzsprung–Russell diagram.

Blue Giants

Supergiants

Brightness

Giants

Main Sequence

White Dwarfs

Spectral Class

| O | B | A | F | G | K | M |

some 11,500,000 km (7,100,000 miles) out.

Hipparchos European astronomy satellite designed to measure precisely the positions, motion and parallax of more than 100,000 stars.

Hipparchus (died 125 BC) Greek astronomer who compiled a star catalogue, and discovered the precession of the equinoxes. He also pioneered the mathematical study of trigonometry.

Hoba meteorite The largest known meteorite, found in 1920 at Hoba West near Grootfontein in southwest Africa. Measuring 2.8 m by 2.4 m (9 ft by 8 ft), it weighs about 59 tonnes.

hold A temporary halt in the countdown before a space launch. It may be a scheduled, or 'built-in' hold, or an unscheduled one due to a check-out problem.

Hooker telescope The famous 100-inch (2.5 m) reflector at the Mt Wilson Observatory, which has been operational since 1917. Edwin Hubble used the Hooker in pioneer observations of the galaxies.

horizon Astronomically, the great circle where the horizontal plane from a point of observation hits the celestial sphere. More generally, the imaginary line where the land or sea meets the sky.

horizon system One system for locating stars, using the horizon as a reference plane. On this system the stellar coordinates are altitude and azimuth.

Horologium The Clock; a minor southern constellation near Eridanus.

horoscope A drawing that shows the relative positions of the Sun, Moon, and the planets at a person's birth. Astrologers use horoscopes to try to make predictions about the future.

Horsehead nebula The prominent dark nebula in the constellation Orion, which looks like a horse's head. Located near the star Zeta Orionis, it is about 1000 light-years away.

HOTOL A horizontal take-off and landing launch vehicle proposed by British Aerospace for space flight in the 1990s. Taking off and landing on a runway, the craft would use a revolutionary air-breathing jet/rocket engine.

Houston The familiar call sign of mission control for US manned space flights. Mission control is located at the Johnson Space Center at Houston in Texas.

Hoyle, Fred (born 1915) English astronomer who, with Gold and Bondi, put forward a steady-state theory of the origin of the universe.

H–R diagram See **Hertzsprung-Russell diagram.**

H-regions Regions of space containing clouds of hydrogen. H–I regions are made up of neutral hydrogen, which emits no visible light but can be detected by its radio emission at the 21 cm wavelength. H–II regions consist of ionized hydrogen gas and are found around very hot stars. They can be detected optically.

Hubble classification A method Edwin Hubble suggested for classifying the galaxies.

Hubble constant The rate at which the speed of recession of the galaxies changes with distance. Its value has been altered drastically since Hubble's time and is presently considered to be about 60 km (35 miles) per second per megaparsec, or some 20 km (12 miles) per second per million light-years.

Hubble, Edwin P. (1889–1953) US astronomer who pioneered study of the galaxies with the Hooker and Hale telescopes. He related the red shift of the spectral lines of galaxies to their recession, and established a relationship between their speed of recession and distance (Hubble's law).

The best-known dark nebula, the Horsehead nebula in Orion.

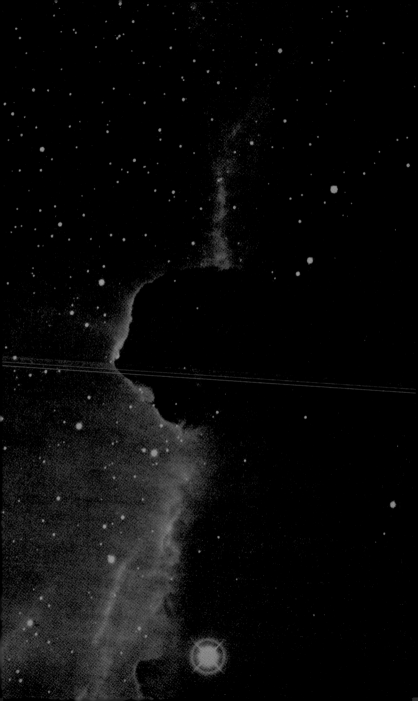

Saturn's moon Iapetus is mysteriously darkened on one side.

Hubble's Law A relationship, first put forward by Hubble in 1929, which states that the red shift in the spectral lines of the light from a galaxy is proportional to its distance. The constant of proportionalilty is known as the Hubble constant.

hunter's moon The full moon that follows the harvest moon.

Hunting Dogs See **Canes Venatici.**

Hyades A famous open star cluster grouped around the star Aldebaran in the constellation Taurus. Visible to the naked eye, it contains about 200 stars.

Hydra The Water Snake; a very long constellation located mainly in southern skies.

hydrogen This is the most abundant element in the universe, and has the simplest atomic structure – one proton and one electron. It makes up most of the matter in stars and is also found scattered between the stars and in nebulae. Nuclear fusion reactions between hydrogen atoms (or rather, protons) provide the energy that keeps the stars shining (see **proton-proton reaction**).

Hydrus The Little Water Snake; or Sea Serpent; a small southern constellation.

hypergolic propellants Those that ignite spontaneously when they mix. A common combination, used for the space shuttle's orbital manoeuvring system engines, mixes nitrogen tetroxide as oxidizer with a form of hydrazine as fuel.

Hyperion One of Saturn's moons, located 1,500,000 km (900,000 miles) out. It is noticeably elongated, measuring some 300 km (200 miles) long, and has an old surface.

I

IAF International Astronautical Federation; an international group of societies, inaugurated in 1951, devoted to promoting the exploration of space.

Iapetus One of the outer moons of Saturn. Orbiting some 3,500,000 km (2,200,000 miles) out, it is about 1400 km (900 miles) across. Interestingly, part of its surface is very dark, part is bright.

IAU The International Astronomical Union; an association of the world's astronomers, founded in Brussels in 1919.

Icarus An asteroid that strays outside the orbit of Mercury and occasionally comes within a few million kilometres of Earth.

ICBM Intercontinental ballistic missile.

Indian See **Indus.**

Indus The Indian; a small southern constellation close to Pavo.

inertial guidance The main means of guiding rockets and spacecraft. In an inertial guidance unit, accelerometers and gyroscopes on three axes sense any change in speed and direction and feed the information to a computer. This computer is programmed with the desired speed and direction at any time. A difference between the actual and desired data prompts the computer to instruct rocket motors to act to correct the errors.

inertial upper stage An additional rocket stage, attached to spacecraft launched by the shuttle which have to be boosted into high orbit.

inferior conjunction See **conjunction.**

A Minuteman ICBM in its underground silo. It is a three-stage rocket missile, which has a range of up to 13,000 km (8000 miles).

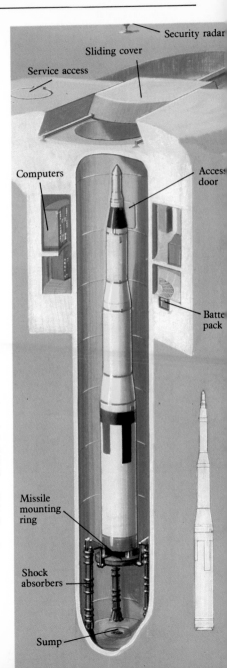

Security radar
Sliding cover
Service access
Computers
Access door
Battery pack
Missile mounting ring
Shock absorbers
Sump

inferior planet Mercury or Venus, whose orbits lie within that of the Earth.

infrared astronomy Astronomy conducted at infrared wavelengths. The atmosphere lets through certain near-infrared wavelengths, which can then be studied from the Earth at locations such as Mauna Kea (the United Kingdom Infrared Telescope, UKIRT) in Hawaii. Far-infrared radiation is absorbed by the atmosphere and thus must be studied from orbit by satellites such as IRAS.

Inmarsat The International Marine Satellite Organization, a body that supports a network of communications satellites for communicating with ships at sea.

inner planets The small rocky planets that orbit relatively close to the Sun – Mercury, Venus, Earth and Mars. They are very much smaller than the four giant outer planets, which are great globes of gas.

Insat An Indian satellite; it combines the roles of direct broadcasting, communications and weather survey.

Intelsat The International Telecommunications Satellite Organization, an international body of some 110 countries that finance a network of communications satellites in geostationary orbit above the Atlantic, Pacific and Indian Oceans.

Intelsat V A powerful communications satellite launched by Intelsat, which can handle some 12,000–15,000 simultaneous telephone conversations, plus two television channels. Intelsat VI (from 1986) has twice this capacity.

interferometer An instrument that uses the interference effects of waves (eg light or radio) for measurement. Stellar interferometers are used, for example, to separate close double stars. Radio interferometry is used in radio astronomy to increase the resolving power of radio telescopes.

intergalactic Between the galaxies.

International Astronautical Federation See **IAF**.

International Astronomical

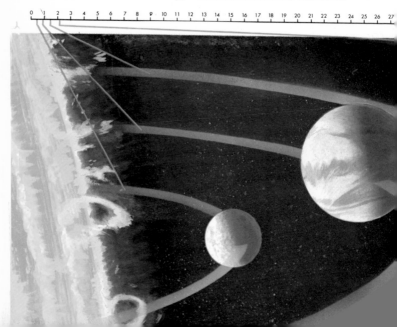

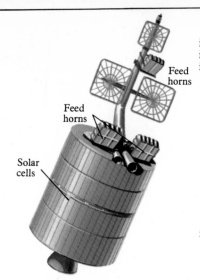

Feed horns

Feed horns

Solar cells

Above: The Intelsat IVA satellite, a communications satellite with some 6000 telephone circuits.
Below: The four inner planets of the solar system.

Union See **IAU**.

interplanetary Between the planets.

interstellar Between the stars.

interstellar matter Gas and dust between the stars in the galaxies. The gas is mainly hydrogen, which may be neutral or ionized (see **H-regions**). The hydrogen regions can be mapped by radio astronomy, wich also reveals a surprising abundance of other molecules,. many of them organic, or carbon-based. They include hydrogen cyanide (HCN), alcohol (C_2H_5OH), ammonia (NH_3), formaldehyde ($HCHO$), ether (CH_3OCH_3), hydrogen sulphide (H_2S) and water (H_2O).

inverse square law A law that applies to the brightness, or intensity of light – starlight, for example. The intensity of a light source varies as the inverse square of the distance at which the source is observed. So the brightness of a star 20 light-years away will only be one-quarter ($1/2^2 = 1/4$) what it is at a distance of 10 light-years.

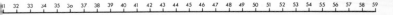

31 32 33 34 35 3o 37 38 39 40 41 42 43 44 45 46 47 48 49 50 51 52 53 54 55 56 57 58 59

Distance in 100s of millions of kilometres

Io Jupiter's third largest moon, which the Voyager probes found to be volcanically active. It measures 3630 km (2255 miles) across, and orbits at a distance of 413,000 km (260,000 miles). It is a vivid orange yellow in colour, due to the volcanic activity, which spews molten sulphur, not molten rock, onto the surface.

ion An electrically charged atom or group of atoms, which has lost electrons. The matter in stars and nebulae is generally ionized and exists in the form of a so-called plasma.

ionosphere A region in the Earth's atmosphere which consists mainly of ionized atoms. It extends from about 55 km (35 miles) to the exosphere, some 500 km (300 miles) high. It contains several distinct layers, which are useful in reflecting radio waves for long-distance communications.

ion rocket An advanced rocket which accelerates ions to form a propulsive exhaust jet. Prototype ion rockets have already been successfully tested. Operational rockets would be large, having a vast array of solar panels to generate a powerful electric field. This would be used to accelerate ions of caesium or mercury to high speed.

IR Infrared.

IRAS This highly successful US/Dutch/English infrared astronomy satellite was launched in 1983. During its 10-month lifetime it spotted regions where stars are being born, and discovered evidence of planetary systems around several stars, including Vega.

IRBM Intermediate range ballistic missile.

irregular galaxy One that lacks structure, such as the Large and Small Magellanic Clouds.

Isaac Newton Telescope A reflecting telescope once located at the Royal Greenwich Observatory at Herstmonceux in Sussex, but now relocated at the Northern Hemisphere Observatory at La Palma in the Canary Islands. It now boasts a slightly larger mirror than before, 254 cm (100 inches).

ISAS The Japanese Institute of Space and Aeronautical Science, which has a launch site at Kagoshima.

Ishtar Terra One of the two main highland regions of Venus, mapped by radar from orbit. Larger than the United States and located in the northern hemisphere, it has the highest mountain range on the planet, Maxwell Montes.

ISRO The Indian Space Research Organization, which has its main launch site, called SHAR, on the island of Sriharikota, north of Madras.

J

Jansky, Karl (1905–1950)
US electrical engineer who first identified persistent radio interference in radio receivers as coming from the heavens. He thus pioneered radio astronomy.

Janus An inner 'moon' of Saturn discovered in 1966. When the Voyager probes visited the planet in 1980 and 1981, a pair of tiny satellites was found in a similar orbit, and are now termed co-orbitals.

Jeans, James Hopwood (1877–1946) English astronomer noted for his theories on the origin of the solar system. He considered the planets to have been formed from gas torn from the Sun by a passing star.

Jet Propulsion Laboratory (JPL) NASA facility at Pasadena, near Los Angeles in California, which is staffed

Opposite: The Voyager 1 probe spotted a volcano erupting on the limb of Jupiter's moon Io. It is the only body in the solar system besides the Earth that is known to be volcanically active.

Right: The infra-red astronomy satellite IRAS, pictured in orbit travelling over Europe.

and managed by the California Institute of Technology. JPL is noted particularly for its support of deep space missions to Jupiter and beyond, and it operates the Deep Space Network of tracking stations.

jet stream High-speed wind belts coursing through the atmosphere. They are present in the atmospheres of Jupiter and Saturn as well as Earth's atmosphere.

jettison To discard, particularly a rocket stage after its fuel has run out.

Jewel Box The magnificent open cluster in the small far southern constellation Crux. Identified as Kappa Crucis, it includes a variety of different coloured stars.

Jodrell Bank Observatory One of Britain's leading radio observatories, located in Cheshire and founded in 1949. Its main instrument is a fully steerable 76-m (250-ft) dish (called the Mark 1A), which became operational in 1957, just in time to track the world's first artificial satellite, Sputnik 1.

Johnson Space Center A NASA establishment near Houston, Texas, which is the main communications centre, or mission control, for US manned space flights. Its call sign, 'Houston', first became familiar during the Apollo missions. Johnson is also a major astronaut training centre and spacecraft test facility.

Julian calendar A calendar devised by Julius Caesar in 46 BC, which included three years of 365 days and a leap year of 366 days every fourth year. Because the year is fractionally longer than 365¼ days, errors crept in over the centuries, which were eventually corrected by Pope Gregory XIII, see **Gregorian calendar.**

Voyager took this picture of Jupiter, showing on the left the Great Red Spot, and on the right the moon Io.

Iron silicate core

Red Spot

Jupiter By far the largest planet in the solar system, being more than twice as massive as all the others put together. Some 143,000 km (88,700 miles) in diameter, it lies on average 780 million km (480 million miles) from the Sun. It has the fastest axial rotation of all the planets, 9 hours 50 minutes. It is mainly a gaseous planet of hydrogen and helium, and its thick atmosphere is marked with parallel bands, which are circulating clouds. Its most prominent feature, though, is the Great Red Spot, a gigantic storm centre. Beneath the atmosphere there is thought to be a vast ocean of liquid hydrogen and beneath that a layer of liquid metallic hydrogen. There may be a small rocky core. Jupiter has no fewer than 16 moons, including the four large Galilean moons Ganymede, Callisto, Io and Europa, which are visible with binoculars.

Under Jupiter's thick atmosphere there probably exist layers of liquid hydrogen and liquid metallic hydrogen.

K

Kagoshima A major launch site for Japanese rockets, operated by the Institute of Space and Aeronautical Science.

Kaliningrad Site of Russia's mission control centre, on the Baltic Sea coast west of Moscow.

Kapteyn, Jacobus Cornelius (1851–1922) Dutch astronomer noted for his discovery in 1904 of two preferential directions for stellar motion, known as star streaming.

Keel See **Carina**.

kelvin The unit on the absolute scale of temperature, symbol K.

Kennedy Space Center This is the world's foremost spaceport, near Cape Canaveral in Florida. It is the main launch and landing site for the American space shuttle, which uses facilities such as the Vehicle Assembly Building and launch pads at Complex 39 built for the Apollo Moon landing project.

Kepler, Johannes (1571–1630) German astronomer who was one-time assistant of Tycho Brahe, and who brought order to planetary motions with his three famous laws. He published the first two laws in 1609, the third 10 years later.

Kepler's laws Fundamental laws of planetary motion first stated by Kepler. The first law – the planets have elliptical orbits around the Sun. The second law – the radius vector (an imaginary line joining the planet and the Sun) sweeps out equal areas in equal times. The third law – the square of the period of a planet in its orbit is directly proportional to the cube of its mean distance from the Sun.

Kepler's star A supernova in the constellation Ophiuchus, witnessed by Kepler in 1604.

kiloparsec One thousand parsecs.

Kirchhoff, Gustav (1824–1887) German physicist who, working with

Above: An aerial view of the Kennedy Space Center in Florida. It shows the massive Vehicle Assembly Building and beyond it the shuttle landing runway.

Below: Johannes Kepler was the first to realise that the planets travel in ellipses around the Sun.

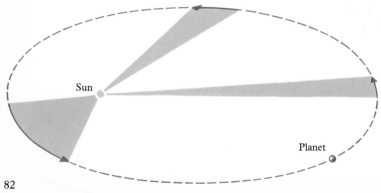

Robert Bunsen (1811–1899), carried out pioneering work in spectroscopy. They found that every element gives off characteristic radiation and gives rise to a characteristic spectrum of bright lines – a bright-line, or emission spectrum. Furthermore, each element is also capable of absorbing radiation identical to that it can give off. This explains the characteristic dark-line, or absorption spectrum of sunlight.

Kirkwood gaps Regions in the asteroid belt where asteroids do not orbit, named after the US astronomer Daniel Kirkwood (1814–1895).

Kitt Park National Observatory One of the world's best equipped observatories, located on Kitt Peak mountaintop, some 80 km (50 miles) south-west of Tucson in Arizona. The 2064-m (6770-ft) high desert site gives perfect viewing conditions for practically every night of the year. Major instruments include the 4-m (157-inch) Mayall reflector and the McMath solar telescope.

Kizim, Leonid (born 1941) One of the three Russian cosmonauts (with Atkov and Solovyov) who established a new space duration record in 1984 by remaining 237 days in Salyut 7.

Komarov, Vladimir Mikhaylovich (1927–1967) Russian cosmonaut who was killed while returning from his second space flight, in Soyuz 1. The re-entry capsule apparently became entangled in its braking parachute high in the atmosphere.

Kosmolyot A name often given to the Russian space shuttle, which is now presumed to be at an advanced stage of development.

Kourou A town in French Guiana, in north-east South America, near the main launching site of Ariane, the launch vehicle of the European Space Agency.

K stars Stars of spectral type K, which are orange and have a surface temperature of about 5000°C.

Kuiper Airborne Observatory
A flying astronomical observatory, being a converted Starlifter transport plane carrying a 0.9 m (3 ft) telescope designed for infrared observations at great height. It is named after a leading Dutch-American planetary astronomer Gerard P. Kuiper (born 1905).

L

Lacerta The Lizard; a small and insignificant northern constellation near Cepheus.

Lagoon nebula (M8) A fine bright nebula in the constellation Sagittarius, just visible to the naked eye. Some 60 light-years across, it lies about 6500 light-years away.

Lagrangian points Gravitationally stable regions in space between two bodies exerting gravitational attraction. The Lagrangian point L5, in the Earth-Moon system, is a region favoured for the site of future space colonies.

Laika The dog that became the world's first space traveller on 3 November 1957, when it was launched into orbit in Sputnik 2.

Landsat US Earth-survey satellite, formerly known as ERTS (Earth Resources Technology Satellite). The first Landsat was launched in 1972. The latest one, Landsat 5, carries advanced scanning equipment known as the thematic mapper, which gives high-resolution images. Landsats scan the Earth's surface at a number of visible and infrared wavelengths. Landsat data are converted into false-colour images which reveal surface detail normally invisible.

Langley Research Center
NASA facility in Hampton, Virginia,

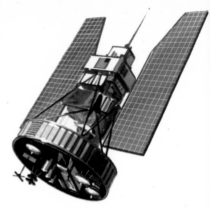

Above: The design for the first three Landsat satellites.

Right: A Landsat picture of the region around Minneapolis-St Paul. The urban areas show up as blue-grey, while vegetation in the countryside shows up as red.

mainly concerned with advanced flight research. But it also has responsibilities for the Scout launch vehicle programme, and managed the Viking Mars landing project in 1976.

La Palma Observatory
See **Northern Hemisphere Observatory.**

Laplace, Pierre Simon de (1749-1827) French mathematician and astronomer who was a pioneer of celestial mechanics, and in 1796 author of the celebrated nebular hypothesis on the origin of the solar system.

Large Magellanic Cloud Nubecula Major; the closer and larger of the pair of naked-eye galaxies visible in far southern skies. Measuring some 30,000 light-years across and located in the constellation Dorado at a distance of some 170,000 light-years, it is classed as a dwarf irregular.

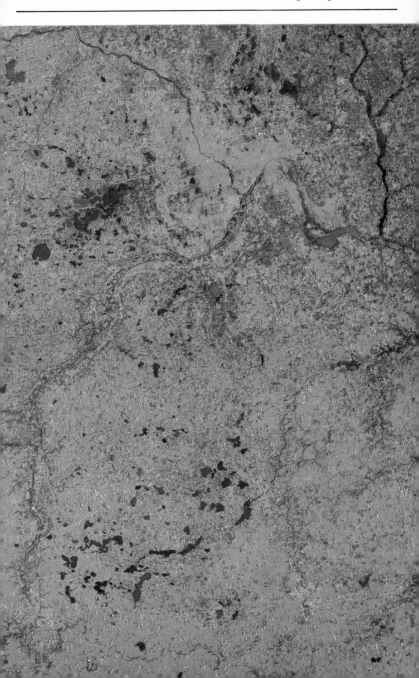

Leasat, launched by the shuttle.

Last Quarter The last of the Moon's four phases, when half the surface is illuminated.

latitude, celestial
See **right ascension.**

launch vehicle A rocket unit for launching a spacecraft into orbit or beyond. To achieve a suitable power-to-weight ratio, it has to be a multistage rocket. Conventional launch vehicles such as Delta and Ariane usually have three stages. The space shuttle has two.

launch window A period of time during which a spacecraft can be launched so as to achieve its desired orbit or trajectory. For all spacecraft there is usually a daily launch window. For deep space probes there is an additional launch window every so many days, months or even years when launchings to the planets are favourable.

LDEF See **Long Duration Exposure Facility.**

leap second An adjustment of one second in time periodically made, usually on 31 December, to allow for the gradual slowing down of the Earth's rotation.

Leap year A year that contains not the usual 365 days, but 366 days and has 29 days in February. Normally, every fourth year is a leap year, when the date year is divisible by four. This is to account for the solar year, which is 365¼ days long. For near-perfect accuracy to solar time, century years are deemed leap years only when they are divisible by four hundred. So the year 2000 will be a leap year.

Leasat A Hughes communications satellite, launched by the space shuttle and leased to the US Navy. It measures some 6 m (20 ft) long and weighs nearly 7 tonnes.

Leavitt, Henrietta S. (1868–1921) US astronomer at Harvard Observatory who derived in 1912 the period-luminosity law for the Cepheids.

Leda One of Jupiter's moons, less than 15 km (10 miles) across, which orbits at a distance of about 11,000,000 km (7,000,000 miles).

Lemaître, Abbé (1894–1966) Belgian mathematician astronomer who was one of the originators of the Big-Bang theory of the origin of the universe.

Leo The Lion; an easily recognizable constellation of the zodiac. The curved front part looks like, and is named the Sickle. Prominent at the bottom of the Sickle's 'handle' is the beautiful mag 1 star Regulus.

Leo Minor Little Lion; a tiny constellation between Leo and Ursa Major.

Leonids An often brilliant meteor shower, with its radiant in the constellation Leo. It occurs between about 12 and 17 November.

Leonov, Alexei A. (born 1934) Russian cosmonaut who made the world's first spacewalk on 18 March 1965, from the Voshkod 2 spacecraft. He later took part in the 1975 Apollo-Soyuz Test Project with US astronauts.

Lepus The Hare; a small and inconspicuous southern constellation.

Leverrier, Urbain Jean Joseph (1811–1877) French astronomer and mathematician, whose investigations of irregularities in the motion of Uranus led to the discovery of Neptune. John Couch Adams carried out similar work at the same time.

Lewis Research Center A NASA field centre near Cleveland, Ohio, concerned with aeronautics and astronautics research, particularly advanced space propulsion. It manages the Centaur and Titan-Centaur launch vehicles.

Alexei Leonov (bottom) is snapped on the ASTP mission with US astronauts.

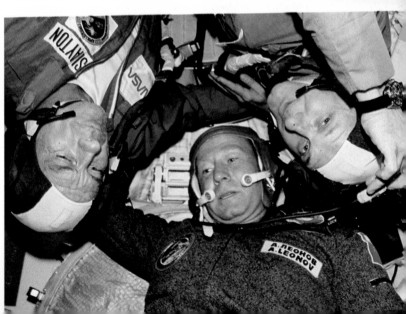

Libra The Scales; an equatorial constellation, seventh in the zodiac.

libration A regular 'nodding' motion of the Moon up and down and from side to side, caused by irregularities in its orbital motion. It enables us to see nearly 60% of the Moon's surface.

Lick Observatory The well-known US observatory on Mt Hamilton, south of San Francisco in California. Operational since 1888, it has a 90-cm (36-inch) refractor, the world's second largest, and a 305-cm (120-inch) reflector.

life-support system A system in a spacecraft that keeps the astronauts alive in space. It provides a constant supply of air to breathe, at the right temperature and pressure.

lift-off The moment a rocket leaves the launch pad.

light The electromagnetic radiation to which our eyes are sensitive. White light, as from the Sun, is a mixture of different wavelengths, or colours. We see them spread out as a spectrum when white light is passed through a spectroscope. The wavelengths vary from 4×10^{-7} metre (violet) to about 7×10^{-7} metre (red).

light curve A graph that records the changes in brightness of a variable star over a period of time.

light, velocity of In a vacuum, 299,793 km (186,282 miles) per second. This is usually taken as 300,000 km (186,000 miles) per second. According to relativity, this speed is the highest that can ever be reached.

light-year A basic unit of measurement used in astronomy. It is the distance light travels in a year, some 9.5×10^{12} km (5.9×10^{12} miles). Professional astronomers now generally use the parsec (= 3.26 light-years) as a unit.

limb The edge of the disc of a heavenly body, such as the Sun or the Moon.

Lion See **Leo**.

Lippershey, Hans (ca 1570–1619) Dutch spectacle-maker who invented the telescope in 1608.

liquid hydrogen A common rocket fuel, used for example in the space shuttle's main engines. It is a cryogenic propellant, at a temperature of $-253°C$.

liquid oxygen The most common rocket oxidizer, a cryogenic pro-

Opposite: The Long Duration Exposure Facility is lifted from the shuttle orbiter's payload bay by the remote manipulator arm.

Left: The first pulsar signals (top) seemed to come from an intelligent source and were dubbed Little Green Men.

pellant at a temperature of $-183°C$.

liquid propellant A rocket propellant in liquid form, such as liquid hydrogen or hydrazine.

Little Bear See **Ursa Minor**.

Little Dipper US name for the constellation Ursa Minor.

Little Dog See **Canis Minor**.

Little Green Men(LGM) The name jokingly given to the first pulsar signals, which at first sight appeared to have come from intelligent aliens.

Little Lion See **Leo Minor**.

Little Water Snake See **Hydrus**.

Lizard See **Lacerta**.

Local Group The small cluster of galaxies to which our own galaxy belongs. Among its other 25 or so members are the Andromeda galaxy and the two Magellanic Clouds. The Group extends over a distance of some 5 million light-years.

Long Duration Exposure Facility (LDEF) A huge US satellite launched in February 1984 and recovered from orbit over a year later. It contained 57 experiments, including study of interplanetary gases, micrometeoroids, cosmic rays and crystal growth.

longitude, celestial See **declination**.

Long March The name of China's main launching rocket, named after an historic march in 1927 led by former Communist party chairman Mao Tse-tung.

long-period variables
  See **Mira stars.**
Looped nebula
  See **Tarantula nebula.**
Lovell, Alfred Charles Bernard
  (born 1913) British radio astronomer who helped found the Nuffield Radio Astronomy Observatory at Jodrell Bank, near Macclesfield, Cheshire.
Lowell, Percival (1855–1916)
  US astronomer noted for his observations of Mars, and champion of the idea of intelligent Martian life. He established the Lowell Observatory at Flagstaff, Arizona, in 1894.
Lox An abbreviation for liquid oxygen.
L-sat The European direct broadcasting satellite built by Britain and Italy and scheduled for launch into geostationary orbit in 1986.
luminosity A star's absolute brightness, measured in magnitudes.
Luna A series of Russian lunar probes. Luna 2 was the first spacecraft to reach the Moon, in 1959. The same year Luna 3 photographed the Moon's hidden side for the first time. In 1966, Luna 3 probes soft-landed on

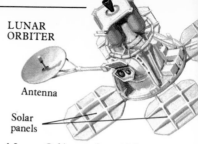

LUNAR
ORBITER

Antenna

Solar
panels

A Lunar Orbiter probe, which mapped the Moon from orbit.

the surface and went into lunar orbit. In 1970, a capsule from Luna 16 returned with a soil sample to Earth, and Luna 17 landed a robot roving vehicle on the Moon, Lunokhod.
lunabase The name for the dark volcanic rock making up the Moon's mare regions.
lunar Relating to the Moon.
lunar calendar A calendar based on the phases of the Moon, a regular cycle of $29\frac{1}{2}$ days. Such calendars do not fit in well with the solar year. Twelve lunar months makes only 354 days. Lunar calendars thus rapidly get out of step with the seasons. The Jewish calendar is lunar. Every few years extra 'leap months' are included

SURVEYOR

Solar
panels

Antenna

Antenna

Landing
leg

Foot
pad

A Surveyor lunar probe, which soft-landed on the Moon, sent back pictures and also analysed the lunar soil.

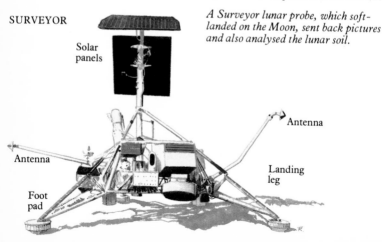

to keep the calendar in step with the solar year.

lunar eclipse An eclipse that occurs when the Moon passes into the Earth's shadow in space. During a lunar eclipse, which always occurs at Full Moon phase, the Moon does not entirely disappear but takes on a coppery hue due to light being refracted by the Earth's atmosphere into the shadow zone. One or two lunar eclipses occur nearly every year.

lunarite The name often given to the light-coloured rock that makes up the highland areas of the Moon.

lunar module (LM) The part of the US Apollo spacecraft that took two astronauts down to the Moon's surface. It was a two-stage construction, the lower, descent module acting as a launch pad for the upper, ascent module when the

Lunokhod was the first wheeled vehicle to travel on the Moon. It was controlled remotely by Russian scientists on Earth.

astronauts left the surface to return to orbit to rendezvous with the parent spacecraft, the Apollo CSM.

lunar orbital rendezvous
The method Apollo astronauts used to land on the Moon and return. They descended to the surface in a separate lunar module, and then returned in it and rendezvoused with the parent craft (the Apollo CSM) in lunar orbit.

Lunar Orbiter A successful series of US lunar probes that thoroughly mapped the surface of the Moon in 1966 and 1967 from lunar orbit.

lunar probe A probe sent to investigate the Moon.

lunar roving vehicle Nicknamed Moon buggy; the transport Apollo astronauts used on the Moon during the last three landing missions (15, 16, 17). A collapsible contraption of aluminium tubing and wire, it was 3 m (10 ft) long and was powered by electric motors on all four wheels, which gave it a top speed of 16 km/h (10 mph).

lunation The period between one New Moon and another, 29½ days.

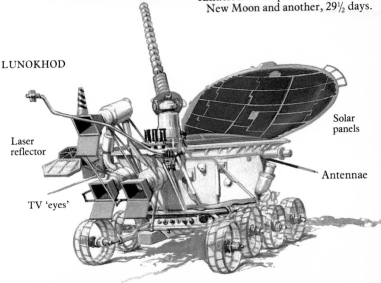

LUNOKHOD

Laser reflector

TV 'eyes'

Solar panels

Antennae

Lunik An alternative name given to early Russian Moon probes, see **Luna.**

Lunokhod The first wheeled vehicle to explore the Moon, launched by Russia. Lunokhod 1 set down on the Moon in Luna 17 in November 1970. Remotely controlled from Earth, it travelled over 10 km (6 miles) during its 10-month life. A second Lunokhod landed in 1973.

Lupus The Wolf; an uninteresting southern constellation.

Lynx A faint northern constellation south-east of Ursa Major.

Lyra The Lyre; a small but brilliant northern constellation close to the Milky Way. Its dominant star is the mag 1 blue-white Vega, one of the stars of the Summer Triangle. Lyra also contains the fascinating Ring nebula (M57).

Lyre See **Lyra.**

Lyrids A meteor shower, with its radiant in the constellation Lyra, which occurs in the third week of April.

Lysithea One of Jupiter's tiny moons, little more than 30 km (20 miles) across. It orbits at a distance of about 11,700,000 km (7,300,000 miles).

M

McCandless, Bruce (born 1938) US astronaut who was the first person to make an untethered spacewalk, on 7 February 1984, during shuttle mission 41–B. He did so while flight-testing the manned manoeuvring unit (MMU).

Mach no A number used to describe very high speeds, as of aircraft and returning spacecraft. It is the ratio of the craft's speed to the local speed of sound. The space shuttle hits the atmosphere travelling at about Mach 25.

McMath solar telescope

James Irwin with the Apollo 15 lunar module and lunar rover.

The world's largest solar telescope, located at Kitt Peak National Observatory in Arizona. A 2-m (6.5 ft) heliostat reflects sunlight down a sloping shaft which extends 90 m (300 ft) below ground. Mirrors then reflect it back up the shaft and into an observation room, where it forms an image of the Sun's disc 76 cm (30 inches) in diameter.

mag See **magnitude.**

Magellanic Clouds The nearest galaxies to our own. See **Large Magellanic Cloud; Small Magellanic Cloud.**

magnetic field A field of influence around a magnetic body, such as a magnet or a heavenly body such as the Sun and the Earth.

magnetic storm A disruption of the Earth's magnetic field and upper atmosphere due to sunspot and flare activity on the Sun. Brilliant auroras occur at these times, and long-distance radio communications are affected.

magnetometer An instrument that measures the strength of a magnetic field. A common instrument on space probes.

magnetosphere An envelope like a bubble around a body in space which marks the boundary of its magnetic field. Because of the solar wind, the magnetosphere of the Earth and other planets is not a sphere, but is more like a teardrop.

The Earth's magnetosphere is distorted by the solar wind.

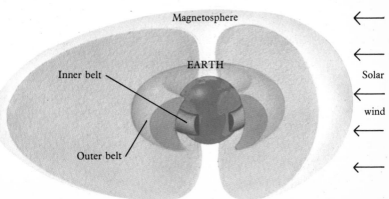

magnitude A measure of a star's brightness. Traditionally the very bright stars were given a magnitude of 1; the faintest detectable to the eye, a magnitude of 6; and the others were divided up proportionately. Today the magnitude scale is precise. A star of the first magnitude (mag 1) is 2.5 times brighter than a star of the second magnitude (mag 2), and so on. The scale is extended to negative values for exceptionally bright stars. It goes beyond 6 for fainter stars. The latest telescopes can spot stars of magnitude 27. The brightness of a star as we see it on Earth is its apparent magnitude. Its **absolute magnitude** is a measure of its true brightness.

main sequence A roughly diagonal region on the Hertzsprung-Russell diagram in which most stars lie.

manned manoeuvring unit (MMU) The 'Buck Rogers' type jet-propelled backpack which shuttle astronauts use to make untethered spacewalks. They move about by firing different combinations of its 24 jet thrusters, which are powered by compressed nitrogen. See **McCandless.**

mantle The part of the Earth's interior between the core and the crust. It is believed to be in a plastic state, allowing the plates of the crust to move, so causing continental drift. It is probably composed of silicates, such as olivine.

mare A flat region of the Moon, consisting of very ancient lava flows. People once thought that mare regions might be seas, and named them after the Latin word for sea, *mare* (plural *maria*).

Mare Crisium Sea of Crises; a noticeably circular mare, some 500 km (300 miles) across near the north-eastern limb.

Marecs A European marine communications satellite, operated by Inmarsat. Launched by Ariane, it operates from geostationary orbit.

Mare Fecunditatis Sea of Fertility; an irregularly shaped mare. It is about 1000 km (600 miles) across, straddling the lunar equator near the eastern limb.

Mare Frigoris Sea of Cold; an irregular elongated mare north of the lunar Alps which leads into Oceanus Procellarum.

Mare Humorum Sea of Moisture; a roughly circular mare to the west of Mare Nubium.

Mare Imbrium Sea of Showers; the largest of the circular maria, with a surface area of some 900,000 sq km (235,000 sq miles). It is ringed by the Carpathians, Apennines, Caucasus, Alps and Jura Mountains.

Mare Nectaris Sea of Nectar; a near circular mare some 500 km (300 miles) across, south of Mare Tranquillitatis.

Mare Nubium Sea of Clouds; an ill-defined mare in the south-west quadrant, merging in the west with Mare Incognitum (Unknown Sea), often considered to be a bay of Oceanus Procellarum.

Mare Serenitatis Sea of Serenity; a circular mare, on the other side of the Caucasus Mountains from Mare Imbrium. It is about 600 km (400 miles) across.

Mare Tranquillitatis Sea of Tranquillity; a fairly irregular shaped mare in the north-east quadrant that merges in the north with Mare Serenitatis. It was the scene of the first Apollo Moon landing in 1969.

Mare Vaporum Sea of Vapours; a small mare just north-east of the centre of the Moon's disc.

Flying a manned manoeuvring unit (MMU), astronaut Robert Stewart becomes a human satellite during the 41-B shuttle mission in February 1984.

Mariner A series of US space probes sent to investigate the nearest planets. Mariner 2, which flew past Venus in 1967, was the first successful probe. Mariner 9 mapped Mars from orbit. Mariner 10 took the first close-up pictures of Mercury in 1973.

Mariner Valley Valles Marineris; the name given to the great 'Grand Canyon' on Mars, which stretches for some 5000 km (3000 miles) near the planet's equator. It is up to 400 km (250 miles) wide and as much as 7 km (4.5 miles) deep.

Mars The fourth planet out from the Sun, with a diameter of 6790 km (4220 miles). It is often called the Red Planet because of its reddish-orange appearance in the sky. It lies on average about 228 million km (142 million miles) from the Sun, around which it circles once in 687 days. Like the Earth, Mars's axis of rotation is tilted in space, giving it seasons. Mars's day is only about 40 minutes longer than our own. Its atmosphere is very slight and consists mainly of carbon dioxide. The polar caps that appear in winter are a mixture of dry ice (frozen carbon dioxide) and water ice. Two outstanding features on Mars are an extinct volcano bigger than Everest, called Olympus Mons, and a great canyon called Mariner Valley.

Marshall Space Flight Center A prime NASA centre for supporting manned space flight, located near Huntsville, Alabama, within the US Army's Redstone Arsenal. It is responsible for the space shuttle propulsion systems, and for development of Spacelab and the space telescope.

mascon A region of exceptionally high density ('mass concentration') on the Moon, detected from lunar orbit by Apollo spacecraft. Most

mascons are found in mare regions.

Maskelyne, Nevil
(1732–1811) English clergyman and astronomer who became the fifth Astronomer Royal in 1765. He is best known as founder of *The Nautical Almanac* in 1767.

mass driver An electromagnetic device proposed for shooting materials mined on the Moon into space for use in space colony construction. It would consist of a series of buckets that would be first accelerated to lunar escape velocity (about 8000 km/h, 5000 mph) and then braked. The materials they hold would be ejected into space where they would be gathered by a suitable catcher.

Below: A Viking probe took this picture of Mars, which shows its four massive extinct volcanoes.

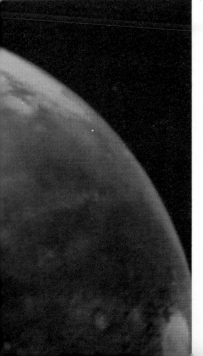

Above: A Martian war machine attacked the Earth in H. G. Wells' book 'The War of the Worlds'.

mass-energy equation An equation developed by Einstein that relates mass and energy, see $E=mc^2$.

Mauna Kea A 4200-m (13,800-ft) peak in Hawaii, site of the United Kingdom Infrared Telescope (UKIRT). With a mirror 3.8 m (12.5 ft) across, it is one of the most powerful of its type in the world.

Maxwell Montes The highest mountain range on Venus, on the 'continent' of Ishtar Terra. It soars to a height of 12 km (7.5 miles).

megaparsec One million parsecs.

Mensa The Table; an inconspicuous southern constellation south of Dorado and containing part of the Large Magellanic Cloud.

97

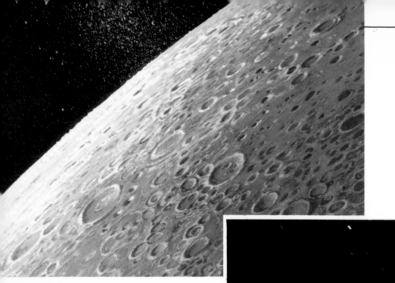

Merbold, Ulf (born 1941) German physicist who became the first European to fly on a US spacecraft when he took part in the Spacelab mission in November 1983.

Mercury The planet closest to the Sun, a barren cratered ball of rock 4850 km (3015 miles) across, on which temperatures soar to more than 500°C. On average it lies some 58,000,000 km (36,000,000 miles) from the Sun, and takes 88 days to make one orbit. It is the second smallest planet in the solar system, after Pluto.

Mercury project The US programme to launch Americans into space. It achieved its first success on 5 May 1961, when Alan B. Shepard made a suborbital flight from Cape Canaveral. But it was 20 February 1962 before John H. Glenn became the first American in orbit. Three more orbital flights followed, ending with Gordon Cooper's 34-hour flight in May 1963.

meridian On Earth, a great circle that passes through the poles perpendicular to the equator. On the celestial sphere, a great circle through the celestial poles, the zenith and the nadir. See **Greenwich meridian**.

MERLIN The multi-element radio-linked interferometer network, a British radio astronomy network that uses the signals from separate dishes dotted around England to synthesize a dish some 130 km (80 miles) across.

meson A fundamental atomic particle found in cosmic rays. Mesons, which include pions and kaons, have a mass smaller than protons and neutrons, and larger than electrons.

Messier number A method of identifying nebulae, star clusters and galaxies by their number in a catalogue compiled by the French astronomer Charles Joseph Messier (1730–1817). He compiled a list of over 100 objects. Among the well known Messier objects are M1 (Crab nebula), M45 (the Pleiades), and M31 (Andromeda galaxy).

meteor Also called shooting star and falling star; a streak of light that results when a piece of rock from outer space burns up as it plunges into the atmosphere. Frictional heating causes it to do so. At certain times of the year there are meteor showers. See also **bolide; meteorite.**

Meteor crater
See **Arizona meteor crater.**

meteorite A piece of rock or metal from outer space that has withstood the frictional heating of Earth's atmosphere and reached the ground intact. Meteorites are made up either of stone (aerolites), metal (siderites), or both (siderolites). The siderites contain mainly iron and nickel. See also **Arizona meteor crater; Hoba meteorite.**

meteoroids Pieces of solid matter present in interplanetary space. They are mostly very small (micrometeoroids). When they enter the Earth's atmosphere they cause meteors. The largest survive to reach the ground as meteorites.

meteor shower A 'rain' of meteors that appear to come from a certain point in the sky, the radiant. Meteor showers occur regularly year by year. They are often associated with the present or past orbits of comets.

Top left: The surface of Mercury is almost entirely covered in craters.

Left: The streak of a meteor in the night sky.

Below: The huge meteorite crater in Arizona, which measures over 1200 metres (4000 ft) across.

Meteosat A series of European weather satellites, in stationary orbit above West Africa. Meteosat 1 was launched in 1977, Meteosat 2 in 1981. The satellites carry radiometers operating at three wavelengths, in the visible, infrared and water-vapour regions of the spectrum.

Metonic cycle A cycle of 19 years, or 235 synodic months, after which time the phases of the Moon occur on the same days of the month. It is named after its discoverer, Meton of Athens (400s BC).

Michelson, Albert A. (1852–1931) Polish-born US physicist who, using interferometry, determined an

accurate value for the velocity of light in 1887, working with Edward W. Morley (1838–1923).

microgravity A term scientists use to describe the condition in orbit which is popularly called weightlessness.

Microscope See **Microscopium**.

Microscopium The Microscope; a tiny inconspicuous southern constellation, near Grus.

Milky Way The milky white band studded with stars that arcs across the heavens. It represents, from our viewpoint, a cross-sectional view of our galaxy.

Milky Way galaxy An alternative name for the Galaxy, the galaxy to which our Sun and all the stars you can see in the sky belong.

Mimas The innermost of the telescopic satellites of Saturn, which circles 185,000 km (115,000 miles) out. Some 400 km (250 miles) across, Mimas has an ancient surface covered in cracks and craters, one 130 km (80 miles) across.

Mimosa (Beta Crucis) A 1st magnitude star in the constellation Crux. It lies some 490 light-years away.

minor planet See **asteroid**.

Mira The star Omicron Ceti in the constellation Cetus. This star is the prototype of the long-period red variables, or Mira stars. Mira varies over roughly 5 magnitudes in about 330 days. It can become as dim as mag 10 or as bright as mag 2 at times.

Miranda The smallest and innermost moon of Uranus. Some 320 km (200 miles) across, it orbits at a distance of 130,000 km (80,000 miles).

Mira stars Long-period variable stars named after the prototype Mira, or Omicron Ceti. They typically have periods of over 100 days. They are thought to be late giant and supergiant stars undergoing brightness changes because of pulsations.

Left: Cloud cover over Europe and Africa, pictured by the weather satellite Meteosat from 35,900 km (22,300 miles).

Below: The Milky Way galaxy as it would appear from space. The Sun (arrowed) lies about 30,000 light-years from the centre.

mission control A centre that takes charge of a space flight. NASA's mission control for manned flights is located at the Johnson Space Center at Houston, Texas, and has the familiar call sign, 'Houston'. Mission control for NASA deep space missions is at the Jet Propulsion Laboratory at Pasadena in California. Russia's mission control centre is located at Kaliningrad on the Baltic coast. Europe's main mission control centre is at Darmstadt, West Germany.

mission specialist The name NASA gives to non-pilot astronauts on the shuttle. These astronauts are responsible for satellite launchings, operation of the remote manipulator system and the day-to-day in-orbit

'house-keeping'. See also **payload specialist**.

Mizar (Zeta Ursae Majoris) The middle star in the handle of the Plough in the constellation Ursa Major. It forms a naked-eye optical double with Alchor. It is itself a visual binary, and in the spectroscope each component is found to be a spectroscopic binary. So Mizar is a multiple four-component star.

MMS See **multi-mission modular spacecraft**.

MMU See **manned manoeuvring unit**.

mobile launch platform The launch platform used by the space shuttle. The shuttle stack is assembled inside the Vehicle Assembly Building (VAB) on the platform, which is then

carried to the launch pad by crawler transporter and set down there. The platform is returned to the VAB after the launch.

module A self-contained unit of a spacecraft.

moldavites Small glassy objects found in Czechoslovakia; forms of tektites.

Molniya This is a series of Russian communications satellites that operate in highly eccentric orbits so as to be above the horizon in Russia for most of the time. Their orbits take them from a low point of some 500 km (310 miles) in the southern hemisphere to a high point of 40,000 km (25,000 miles) in the northern hemisphere.

Monoceros The Unicorn; an equatorial constellation containing many

The operations control room at mission control, Houston, Texas, during a shuttle mission. Flight controllers sit at banks of consoles equipped with video screens and studded with switches.

faint stars, clusters and nebulae.

month Literally the time it takes the Moon to circle the Earth. This can be defined in various ways, depending on the point of reference. The sidereal month of 27.32 (27⅓) days is the time the Moon takes to make one revolution in relation to the stars. The synodic month of 29.53 (29½) days is the time it takes between phases, that is, in relation to the Sun. The arbitrary calendar month varies between 28 and 31 days.

103

Moon The Earth's nearest neighbour in space, its only satellite, and the only other world visited by human beings. It lies on average 384,000 km (239,000 miles) away, and has a diameter of 3476 km (2160 miles). With only about one-eightieth of the Earth's mass, its surface gravity is only one-sixth of that on Earth. The Moon is an airless, barren world of mountains, craters and vast lava plains, or maria (see **mare**). It has a captured rotation, meaning that it keeps the same face towards us as it circles the Earth every 27⅓ days (the sidereal month). It goes through its phases every 29½ days (synodic month). See **phases of the Moon; tides**.

moon A satellite.

Moon buggy
See **lunar roving vehicle**.

morning star Once called Phosphorus; the planet Venus when it appears in the eastern sky before sunrise.

Mount Hopkins Peak in Arizona where the first multiple-mirror telescope is sited.

Mount Palomar Peak in California that is the site of the Palomar Observatory and its famous Hale telescope, operational since 1948.

Mount Wilson Peak in California, some 50 km (30 miles) away from Mt Palomar, that is the home of the Mount Wilson Observatory (1904). Its chief instrument is the 100-inch (254-cm) Hooker telescope, which became operational in 1917.

MSS See **multispectral scanner**.

M stars A class of red stars – dwarfs and giants – whose surface temperature is about 3000°C.

Mullard Radio Astronomy Observatory A world-famous radio astronomy facility near Cambridge and operated by the University. One of the major instruments is the Five-Kilometre Telescope, which consists of eight dishes spaced along a railway track 5 km (3 miles) long. Using the Earth's rotation, it synthesizes a dish 5 km (3 miles) across. See **aperture synthesis**.

multi-mission modular spacecraft (MMS) A standardized NASA spacecraft made up of basic modules (for communications, power, attitude-control, etc), but which can carry different instrument packages. The MMS formed the basis of Solar Max and Landsat 4 and 5.

multiple-mirror telescope (MMT) A novel type of reflector which uses, not a single large mirror to gather starlight, but a number of small mirrors. The first MMT was built on Mount Hopkins in Arizona. It comprises six mirrors 1.8 m (71 inches) across. Computer-controlled, they are equivalent to a single dish 4.5 m (177 inches) across.

Left: The Full Moon is the best time to observe crater rays (streaks radiating from the craters).

Below: Part of the Five Kilometre Telescope at the Mullard Radio Astronomy Observatory.

multiple star A star that consists of more than two components which are held together by their mutual gravitational attraction. The Trapezium, or Theta Orionis is a well-known multiple star in Orion.

multispectral scanner (MSS) One of the imaging instruments carried by Landsat. It scans the Earth at four separate wavebands in the visible and near-infrared region of the spectrum.

multistage rocket An alternative name for **step rocket**.

Murchison meteorite A meteorite that fell at Murchison, Australia, in 1969, in which traces of amino acids were found. The presence of such life-forming compounds suggested to some that ingredients for life may be widely distributed in space.

Musca The Fly; a small southern constellation close to the Southern Cross.

N

N-I, N-II The designations of two of Japan's major rocket launch vehicles.

nadir The point on the celestial sphere vertically below an observer. It is the opposite point to the zenith.

naked eye The unaided eye. Naked-eye observations are those made without the aid of a telescope.

NASA The National Aeronautics and Space Administration, the US agency responsible for the country's programme of space exploration. Founded in 1958, NASA has its headquarters in Washington DC and its main launch site at Cape Canaveral in Florida. At the Cape the manned launches take place at Complex 39 of the Kennedy Space Center, while unmanned launches take place nearby within the confines

of the Cape Canaveral Air Force Base. NASA's mission control is located at the Johnson Space Center at Houston, Texas. Other major NASA centers concerned with space flight include the Marshall Space Flight Center at Huntsville, Alabama, the Jet Propulsion Laboratory at Pasadena, near Los Angeles, in California, and the Goddard Space Flight Center at Greenbelt, Maryland, not far from Washington DC.

NASCOM NASA's communications network.

Right: The aerial of the National Astronomy and Ionosphere Center's Arecibo radio telescope.

Below: NASDA's main launch site on the island of Tanegashima, in the extreme south of Japan.

NASDA Japan's National Space Development Agency, which is responsible for the nation's space effort.

National Astronomy and Ionosphere Center The correct title for the radio observatory at Arecibo in Puerto Rico, which operates the 305-m (1000-ft) radio telescope there. Cornell University manages the facility under the auspices of the National Science Foundation.

National Radio Astronomy Observatory A US observatory near Green Bank, Virginia. Its main instrument is a 91-m (300-ft) dish radio telescope. NRAO also operates the Very Large Array near Socorro, New Mexico.

Nautical Almanac An almanac, established in 1765, published for the benefit of astronomers and navigators, including ephemerides –

107

details about the positions of the heavenly bodies throughout the year. Today such details are found in *The Astronomical Ephemeris*, which in the US is known as *The American Ephemeris and Nautical Almanac*.

Navstar A series of US navigation satellites, which will eventually form a world-wide network known as the global positioning system.

neap tide The low tide that occurs at times of the quarter moon phases, when the gravitational actions of the Sun and Moon oppose one another.

nebula A cloud of dust and gas in space. Bright nebulae shine either because they emit light of their own (emission nebulae) or reflect light from nearby stars. Dark nebulae are dense dust clouds that obscure the light from stars behind them. Planetary nebulae are spheres of expanding gas resulting from the explosion of a central star. Other nebulae resulted from a supernova explosion – they are supernova remnants. The nebulae we see from Earth are within our own galaxy – they are galactic nebulae. Galaxies, which appear nebula-like, were once called extragalactic nebulae.

nebular hypothesis A theory of the origin of the solar system first mooted in the 18th century by Immanuel Kant and Pierre Laplace and now widely accepted by modern astronomers. It supposes that the Sun and planets condensed from a rotating disc which formed when a nebula of gas and dust contracted.

Nemesis The name given to a hypothetical companion star to the Sun. According to a recent theory, it may periodically disturb the Oort cloud of comets in deep space and direct them into the inner solar system in greater numbers than usual. As a result of this, cometary debris rains down on the Earth and causes mass extinctions of animal species (such as the dinosaurs).

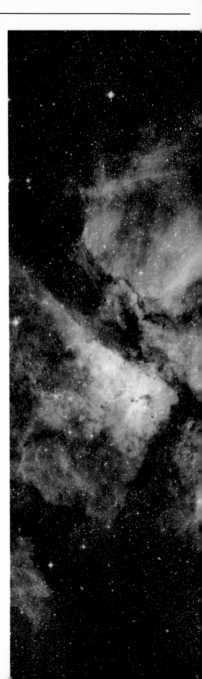

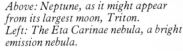

Above: Neptune, as it might appear from its largest moon, Triton.
Left: The Eta Carinae nebula, a bright emission nebula.

Neptune At present and until 1999, the most distant planet in the solar system. Thereafter, Pluto will become the most distant planet again. This anomaly occurs because of Pluto's eccentric orbit. Neptune lies on average about 4500 million km (2800 million miles) from the Sun and takes 165 years to circle round it. Some 50,000 km (31,000 miles) in diameter, it is a gaseous planet like Jupiter, and composed mainly of hydrogen and helium. It has two satellites, Triton and Nereid.

Nereid The outer of the two moons of Neptune, which orbits at a distance of 5,500,000 km (3,500,000 miles). It measures about 900 km (550 miles) across.

NERVA Nuclear engine for rocket vehicle application; a US project of the 1960s for developing a nuclear rocket engine for a proposed manned expedition to Mars. A prototype engine was successfully tested in 1969, but the project was abandoned a few years later.

Net See **Reticulum**.

neutral buoyancy chamber Also called a water immersion facility (WIF); a huge water tank in which astronauts train for spacewalking. They wear specially weighted spacesuits that give them neutral buoyancy, a state in which they neither rise nor sink. It approximates to the condition of weightlessness they will experience in orbit.

neutrino A mysterious fundamental particle that is produced in nuclear reactions, such as those that take place inside the Sun and stars. It has no mass or electric charge and is thus very penetrating, passing readily through matter of any kind.

neutron One of the three commonest fundamental atomic particles (with the proton and electron). Neutrons are present in the nuclei of all atoms except hydrogen. Electrically neutral, they have roughly the same mass as the positively charged proton.

neutron star An incredibly dense body, measuring little more than about 25 km (15 miles) across. It is the collapsed remains of a star that has suffered a supernova explosion. The collapsing process also causes the neutron star to rotate rapidly. As it does so, it gives out radio waves in a directional beam, rather as a lighthouse does. From Earth we see this beam as a pulsating source. This is believed to provide the explanation of the phenomenon of pulsars.

New General Catalogue (NGC) A catalogue of star clusters and nebulae compiled originally by J.L.E. Dreyer in 1888. Such bodies are now usually identified by their NGC number. See also **Messier number.**

New Moon The phase of the Moon

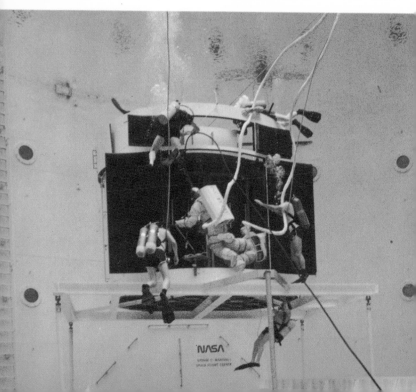

Above: Isaac Newton investigated the spectrum of sunlight and established the laws of gravity. He also built the first reflecting telescope (right).

Opposite: Bruce McCandless and Robert Stewart train in a neutral buoyancy chamber at Huntsville.

when it is in conjunction, that is, when its face is dark. Popularly (but inaccurately) it is called 'New Moon' when a slim crescent first appears.

newton The unit of force in the International System of Units. It is the force that would give a mass of 1 kg an acceleration of 1 m per sec per sec. (1 newton = 0.225 pound.)

Newton, Isaac (1642–1727) English genius who revolutionized mathematics and the sciences. In optics, he investigated the spectrum and built the first reflecting telescope (1671). In mechanics, he established three fundamental laws of motion, of which the third law explains the principle of rocketry. But perhaps his most momentous discovery was the law of gravitation – that every body in the universe attracts every other body. This forms the basis of celestial mechanics.

Newtonian telescope A reflector of the type first built by Newton, which is favoured by most amateur astronomers because it gives a convenient viewing position. In a Newtonian reflector, a curved primary mirror gathers the light and reflects it back up the telescope tube to a plane mirror angled at 45°. This reflects the light into an eyepiece set in the side of the tube.

Newton's third law To every action there is an equal and opposite reaction. This explains how a rocket works. The action of the hot gases escaping (backwards) out of the rocket nozzle is accompanied by an equal and opposite (forwards) reaction that propels the rocket.

NGC See **New General Catalogue**.

night The time in that part of the world where the Sun is beneath the horizon.

Nimbus A series of seven polar-orbiting US weather satellites, the first one of which was launched in 1964. By 1985 only two craft were still operational.

NOAA The National Oceanographic and Atmospheric Administration, a US government organization concerned with monitoring the environment. It operates a network of weather satellites, designated NOAA. These satellites provide most of the cloud pictures seen on television weather forcasts.

noctilucent clouds Luminous clouds that can sometimes be observed high in the sky long after sunset. More than 80 km (50 miles) high they are composed of dust particles, probably from interplanetary space.

nodes The points at which one orbit intersects another. They refer in particular to where the orbit of a planet or other heavenly body cuts the ecliptic. It is called the ascending node when the body is moving north and the descending node when it is moving south.

NORAD The North American Aerospace Defense Command, which keeps a constant radar watch on objects in orbit.

Norma The Rule; an unremarkable southern constellation in the Milky Way north of Triangulum Australe.

North American nebula A bright nebula near Deneb in Cygnus which has roughly the shape of North America.

Northern Cosmodrome A major Russian launch base, located at Plesetsk in the far north of the country near Archangel. It launches communications and weather satellites, and is also a major missile base.

Northern Cross See **Cygnus**.

Northern Crown
See **Corona Borealis**.

Northern Hemisphere Observatory Also called La Palma Observatory; a recently established international observatory at La Palma in the Canary Islands. It is a joint venture of Britain, Denmark, Spain and Sweden. The two main telescopes are British: the refurbished Isaac Newton telescope (2.5 m, 100 inches), and the new William Herschel telescope (4.2 m, 165 inches), third largest in the world.

North Star
An alternative name for **Polaris**.

nova A faint star that suddenly flares up and becomes thousands of times brighter than usual, then it gradually fades back to its normal brightness. It is thought that nova outbursts occur in binary star systems as matter attracted from one star builds up on the other and explodes. Some nova

repeat the process; these are recurrent nova.

N stars Very red stars of low surface temperature.

Nubecular Major
See **Large Magellanic Cloud**.

Nubecular Minor
See **Small Magellanic Cloud**.

nuclear battery A power unit used in space probes, properly named **radioisotope thermoelectric generator**.

nuclear energy Energy given out as heat, light and radiation in nuclear reactions, such as fission and fusion. Nuclear fusion reactions occur in the interior of stars, where light atoms such as hydrogen fuse together. When fusion takes place, a certain amount of matter is converted into energy according to Einstein's equation $E=mc^2$. It is this energy that keeps the stars shining. See **carbon-nitrogen cycle; proton-proton reaction**.

nuclear rocket A rocket that derives its energy from a nuclear reactor. In the rocket a propellant such as hydrogen is heated by the reactor and accelerated to high speed to form a propulsive jet. NASA tested a prototype nuclear rocket called NERVA in 1969.

nucleus The centre of an atom, where most of its mass is concentrated. Nuclei are made up of two main

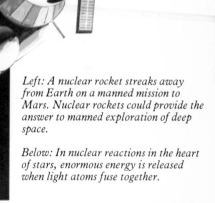

Left: A nuclear rocket streaks away from Earth on a manned mission to Mars. Nuclear rockets could provide the answer to manned exploration of deep space.

Below: In nuclear reactions in the heart of stars, enormous energy is released when light atoms fuse together.

particles, positively charged protons and neutral neutrons, except in the hydrogen atom. Hydrogen, the simplest atom, has a single proton in its nucleus. Circling around the nucleus of the atom is a cloud of negatively charged electrons.

nucleus, comet The solid part in the head of the comet, consisting of ice and dust. It is surrounded by a luminous cloud, or coma of gas and dust.

nutation A slight 'nodding' of the Earth's axis in space, caused by the gravitational attraction of the Moon and the Sun.

O

OAO A series of US orbiting astronomical observatories, which helped pioneer satellite astronomy. Most successful was OAO-3, named Copernicus, which was launched in 1972 and operated for nine years.

Oberon The outermost known moon of Uranus. Orbiting some 590,000 km

(365,000 miles) from the planet, it measures about 900 km (560 miles) across.

Oberth, Herman (born 1894) German who helped found the science of astronautics. He achieved widespread recognition with his book *The Rocket into Interplanetary Space* (1934). He later worked on the development of the V-2 rocket with Wernher von Braun.

objective or object glass; the lens or lens system of a refractor that forms the image (which is further magnified by the eyepiece). The objective's light-gathering power increases as the square of its diameter.

Left: This large refractor at the Allegheny Observatory, Pittsburgh, is used to pinpoint the position of stars accurately.

Below: Kitt Peak Observatory in Arizona, located at an altitude of over 2 km (1¼ miles).

oblateness The amount a body such as a planet deviates from an ideal sphere shape. Planets are oblate because of their rotation, which flattens them at the poles. This is particularly pronounced with the gaseous planets such as Jupiter and Saturn.

observatory A place for observing the stars. Early astronomical observatories were often located near towns (eg Royal Greenwich Observatory). Modern observatories are located in remote regions away from the glare and pollution of urban areas and as high up as possible to be above the obscuring effects of the atmosphere. Among notable modern observatories are Kitt Peak, in Arizona; Siding Spring in New South Wales; and the Northern Hemisphere Observatory on La Palma in the Canary Islands. The last half-century has seen the growth of the radio astronomy observatory. This need not be located high up but it does need to be away from man-made radio noise.

Ocean of Storms See **Oceanus Procellarum**.

Oceanus Procellarum Ocean of Storms; the biggest mare region on the Moon. Occupying much of the western hemisphere, it covers an area of some 5 million square km (2 million square miles).

occulation The cutting off of the light of one heavenly body by another passing in front of it. The occulation of a star by the Moon, for example, provides a method of accurately measuring the position of the Moon.

Octans The Octant; a faint constellation that contains the south celestial pole.

Octant See **Octans**.

ocular Another term for **eyepiece**.

Olber's paradox If there is an infinite number of stars uniformly distributed in the heavens, why isn't the sky blazing with light, instead of being dark at night? This paradox is named after the German astronomer H.W.M. Olbers (1758–1840), who discussed it in the 1820s. His explanations were not valid because of false assumptions. It is believed now that the recession of the galaxies and the red shift of the starlight reduces the brightness.

Olympus Mons Once called Nix Olympica; a gigantic extinct volcano

found in the northern hemisphere of Mars, which is 500 km (300 miles) across at the base and rises to nearly 30 km (20 miles). It is close to three other huge volcanoes on the Tharsis Ridge.

Omega Centauri A bright globular cluster in the constellation Centaurus, visible to the naked eye. It lies at a distance of 1700 light-years and contains over 100,000 stars.

Omega nebula (M17) Also called the Horseshoe nebula; a bright nebula showing opaque dust lanes, in the constellation Sagittarius. It is a source of radio waves. It measures about 27 light-years across.

O'Neill cylinders A concept for a space colony advanced by US physicist Gerard K. O'Neill. The cylinders, which would contain an Earth-type landscape, would be manufactured from materials mined on the Moon. The most ambitious scheme envisages cylinders up to 30 km (20 miles) long and housing 20,000 people. See also **mass driver**.

Oort cloud A region on the outer edges of the solar system which contains a mass of cometary material. Comets journey into the inner solar system when they have been knocked out of the cloud by some disturbing influence. See **Nemesis**.

open cluster A loose group of perhaps several hundred young stars that move through space together. Best known of the clusters visible to the naked eye are the Pleiades and the Hyades (both in Taurus) and Praesepe, the Beehive (in Cancer).

open universe A concept of the universe which says that it will continue its present expansion for ever. Only if there is sufficient matter present in the universe will gravitation eventually halt the expansion and reverse it. See **closed universe**.

Ophiuchus The Serpent-Bearer; a constellation that straddles the celestial equator between the two halves of Serpens (Caput and Cauda, or Head and Tail). It contains two fine globular clusters (M10 and M12), visible with binoculars. **Kepler's star** appeared in the south-east corner of this constellation.

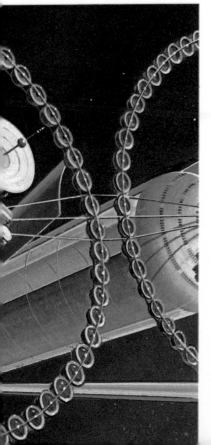

A pair of O'Neill cylinders, home for thousands of Earth colonists next century. The cylinders would be rotated to create an artificial gravity. Sunlight would be reflected inside by long mirrors, creating an ideal climate.

117

opposition The position of a planet when it lies on the opposite side of the Earth from the Sun, and in line with them. It is usually the best time for making observations of the planet.

optical astronomy Astronomy in visible light, or through the 'optical window' of the Earth's atmosphere.

optical double A double star whose components are not physically associated. They appear together because they just happen to lie in the same direction in space. Alchor and Mizar are a well-known optical double.

orbit The path through space of one body around another under the influence of gravity. More precisely, the two bodies orbit around a common centre of gravity, or barycentre.

Orbita A communications network of satellites and earth stations operated by Russia. The main satellites used are Molniya.

orbital manoeuvring vehicle A craft designed for use in orbit when space stations become operational. These vehicles will be used for visiting, repairing and recovering satellites.

orbital period The time it takes a satellite to make one orbit of the Earth. At a height of 300 km (200 miles) the period is about 90 minutes. At a height of 35,900 km (22,300 miles) the period is exactly 24 hours, see **geostationary orbit**.

orbital velocity The speed a satellite requires to remain in orbit around the Earth at a given altitude. The minimum speed a body must acquire to go into a low orbit above the Earth is about 28,000 km/h (17,500 mph). Although bodies in orbit appear to be weightless, they are still in fact subject to the Earth's gravity. They are in a state of free fall. They must be given a much higher speed to escape entirely from the Earth. See **escape velocity**.

orbiter The main item of hardware in the space shuttle system. It is the part that carries the crew and payload. It

Below: The world's first re-usable space vehicle, the American space shuttle. It is pictured here with one of its special payloads, the space laboratory Spacelab.

Top right: At launch, the orbiter sits on its external fuel tank. Twin solid rocket boosters provide extra take-off thrust for the first two minutes of flight.

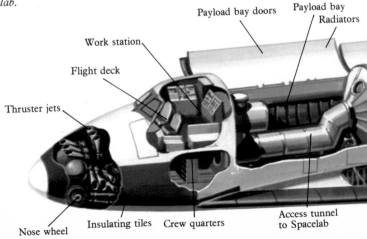

Payload bay doors

Payload bay Radiators

Work station

Flight deck

Thruster jets

Nose wheel   Insulating tiles   Crew quarters

Access tunnel to Spacelab

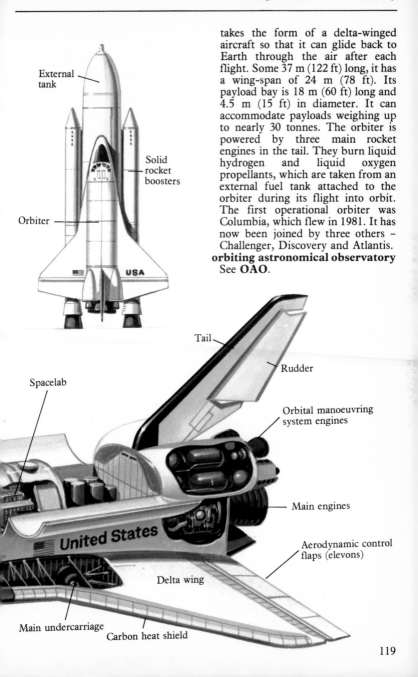

External tank

Solid rocket boosters

Orbiter

USA

takes the form of a delta-winged aircraft so that it can glide back to Earth through the air after each flight. Some 37 m (122 ft) long, it has a wing-span of 24 m (78 ft). Its payload bay is 18 m (60 ft) long and 4.5 m (15 ft) in diameter. It can accommodate payloads weighing up to nearly 30 tonnes. The orbiter is powered by three main rocket engines in the tail. They burn liquid hydrogen and liquid oxygen propellants, which are taken from an external fuel tank attached to the orbiter during its flight into orbit. The first operational orbiter was Columbia, which flew in 1981. It has now been joined by three others – Challenger, Discovery and Atlantis.

orbiting astronomical observatory See **OAO**.

Tail

Rudder

Spacelab

Orbital manoeuvring system engines

Main engines

Aerodynamic control flaps (elevons)

United States

Delta wing

Main undercarriage

Carbon heat shield

119

Orion One of the finest and most easily recognizable constellations, which straddles the celestial equator. Its two brightest stars, at opposite corners of its roughly quadrilateral shape, are the reddish-orange Betelgeuse in the north and the blue-white Rigel in the south. Three bright stars form Orion's belt in the middle, and south of them is the Orion nebula, easily visible to the naked eye.

Orionids A meteor shower that occurs in the third week of October each year, with its radiant in Orion. It is associated with debris from the orbit of Halley's comet.

Orion nebula (M42) The Great Nebula in Orion; the most distinctive naked-eye nebula, in the constellation Orion. It is located beneath Orion's Belt. The nebula glows brightly because of the many hot stars embedded within it, including the magnificent multiple star Theta Orionis, the Trapezium. Some 16 light-years across, the Orion Nebula lies about 1500 light-years away.

Above: The Orion nebula.
Below: The outer planets of the solar system, separated from the inner ones by the asteroid belt.

Orion's Belt Three bright stars in the centre of the roughly quadrilateral shape of the constellation Orion.

orrery A device that demonstrates the motion of the bodies in the solar system. An Englishman, George Graham, designed the first one in about 1715 and named it after the Earl of Orrery. Such a device was once called a planetarium.

oscillating universe A concept about the evolution of the universe, which suggests that it alternately expands from a Big Bang (as it is doing at present); then starts to contract into a Big Crunch; which in turn blasts apart again with another Big Bang; to begin another expansion phase. The process is repeated continuously.

O stars Stars of spectral type O. They are greenish-white stars with surface temperatures of 50,000°C or more.

OTRAG Orbital Transport und Raketen-Aktiengesellschaft; a German company set up to develop and market standard rocket launch vehicles. It fired the first test rockets from Zaire in 1977.

outer planets The planets beyond Mars going away from the Sun – Jupiter, Saturn, Uranus, Neptune and Pluto.

oxidizer A rocket propellant that provides oxygen, such as liquid oxygen and nitrogen tetroxide.

Ozma project A SETI (search for extraterrestrial intelligence) project in which US radio astronomer Frank Drake 'listened' in 1960 for radio signals from space that could come from an alien civilization. He used the 26-m (85-ft) radio telescope at Green Bank, West Virginia, and directed it towards two nearby stars similar to the Sun, Tau Ceti and Epsilon Eridani. He listened for some months, but with no success. A more ambitious SETI programme, Ozma II, was conducted at Green Bank between 1972 and 1976 by P. Palmer and B. Zuckerman. They used the 91-m and 43-m (300- and 140-ft) aerials to scan over 650 of the nearest Sun-like stars, but again picked up nothing that could be regarded as 'intelligent'.

31 32 33 34 35 36 37 38 39 40 41 42 43 44 45 46 47 48 49 50 51 52 53 54 55 56 57 58 59

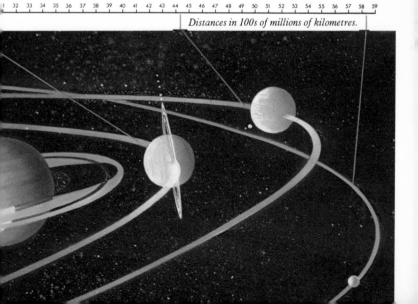

Distances in 100s of millions of kilometres.

ozone layer A layer that exists in the atmosphere at an altitude of between about 10 and 50 km (6 and 30 miles). It is essential to life on Earth because it filters out dangerous ultraviolet radiation in sunlight. Ozone is an isotope of oxygen with three, not two, atoms in its molecules.

P

Painter See **Pictor**.
Palapa A series of Indonesian communications satellites launched by NASA.
Pallas The second largest asteroid, which is thought to be about 610 km (380 miles) in diameter, and the second to be discovered, by H.W.M. Olbers, in 1802.
Palomar Observatory
See **Mount Palomar**.
PAM Payload assist module; an additional rocket stage attached to satellites to boost them into geostationary orbit. PAMs are used on satellites launched by the shuttle or by conventional rockets, such as Delta.

Below: Illustration of the principle of parallax. When a nearby star is sighted from two extremes of the Earth's orbit, it appears to change position. By simple geometry, the distance to the star can then be found.

Above: Launch of a satellite from the shuttle. The attached PAM will later fire to boost it into geostationary orbit.

Top right: The 64-m (210-ft) dish of the Parkes radio telescope in New South Wales.

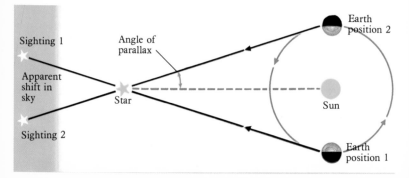

Sighting 1

Apparent shift in sky

Sighting 2

Angle of parallax

Star

Sun

Earth position 2

Earth position 1

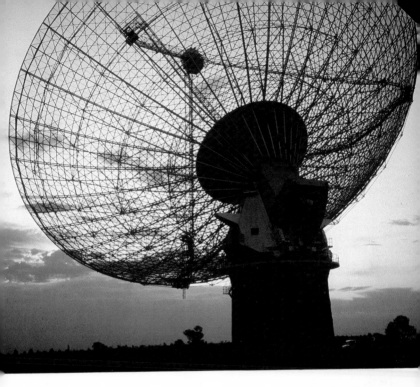

Pangaea The Earth's original supercontinent, which eventually divided up into the present land masses because of continental drift.

parabolic Having the shape of a parabola. Most reflecting telescope mirrors and radio telescope dishes have this shape. This ensures that a parallel beam of light or radio waves is brought to a sharp point focus.

parallax The apparent change in position of a nearby object against a distant background, when it is observed from a different viewpoint. Parallax provides a method of measuring the distance to some of the nearer stars. To determine stellar parallax, a star is viewed from opposite ends of the Earth's orbit. From the apparent change in position of the star, its distance can be estimated. The parallax of a star is defined as the angle subtended at the star by the radius of the Earth's orbit. Even the nearest star, Proxima Centauri, has a parallax of only 0.76 seconds of arc.

Parkes radio telescope A 64-m (210-ft) dish at Parkes Observatory at Coonabarabran in New South Wales, noted for its work on quasars.

parking orbit A temporary orbit into which a spacecraft is launched so that it can be checked out and then boosted accurately into its desired orbit or trajectory.

parsec An astronomical unit used to measure stellar distances. Its name is a contraction of 'parallax' and 'second'. It is defined as the distance at which the radius of the Earth's orbit subtends an angle of 1 second (that is, the parallax is 1 second). 1 parsec = 3.26 light-years.

partial eclipse A solar eclipse in which the Moon does not completely cover the Sun's disc.

Pasiphae One of the outer moons of Jupiter, orbiting more than 23,000,000 km (14,000,000 miles) away. It can be no more than about 50 km (30 miles) across.

path of totality The path of the shadow on the Earth during a total solar eclipse. Even in the most favourable eclipse the path is only about 250 km (150 miles) across.

Pavo The Peacock; a southern constellation near Triangulum Australe. It contains a Cepheid variable (Kappa Pavonis), whose brightness changes are visible to the naked eye.

payload The cargo a space launch vehicle carries.

payload bay The cargo-carrying hold of the space shuttle orbiter, which can accommodate objects up to 18 m (60 ft) long and 4.6 m (15 ft) across.

payload specialist A shuttle astronaut, not necessarily a professional, who has special responsibility for a particular payload.

Peacock See **Pavo**.

Pegasus The Flying Horse; an unmistakable northern constellation, recognized by its 'square'. In the north it extends into Andromeda.

penumbra The partial shadow cast by a planet or a Moon around a central total shadow, or umbra. When the penumbra of the Moon falls on the Earth, a partial solar eclipse occurs. The penumbra of a sunspot is the light area around the dark centre.

perigee The point in the orbit of a satellite or the Moon when it is closest to the Earth.

perihelion The point of closest approach to the Sun in the orbit of a planet or comet.

period The characteristic time interval in a regularly occurring cycle. The orbital period of a satellite is the time it takes to make one orbit of the Earth. The period of a variable star is the time between one maximum (or minimum) brightness and the next.

periodic comet A comet with a relatively short period, such as Encke's (3.3 years) and Halley's (76 years). They are usually designated with a prefix P, such as P/Halley.

period-luminosity law An invaluable law for estimating stellar distances, discovered by Henrietta Leavitt in 1912. She found that there is a precise relationship between the period of the brightness variation of Cepheids and their absolute luminosity (or magnitude) – the longer the period, the greater is the luminosity.

Perseus A northern constellation straddling the Milky Way. It is rich in star fields, and features the star Algol, which was the first eclipsing binary to be recognized. It also contains a fine double cluster of supergiant stars, often called the Sword Handle.

Perseids An often spectacular meteor shower that occurs from late July to mid-August, with the radiant in Perseus.

perturbation A deviation in the orbital motion of a heavenly body because of the gravitational attraction of other bodies (which may not always be evident). Observed perturbations in the orbit of Uranus led to the discovery of Neptune.

phase The part of the surface of the Moon or a planet illuminated by the Sun, which alters when the relative position of the bodies changes.

This picture of the shuttle orbiter Challenger was taken by a camera on Bruce McCandless's manned manoeuvring unit as he made the first untethered spacewalk in February 1984. It shows satellite launching pods inside the open payload bay.

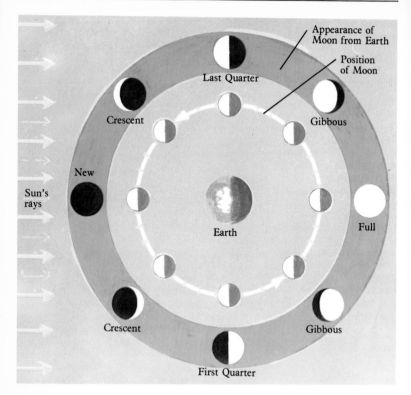

phases of the Moon The change in shape of the Moon's illuminated surface as it orbits the Earth. There are four main phases. At New Moon, the face is dark because the Sun is illuminating only the hidden side. At First Quarter and Last Quarter, half the face is illuminated. At Full Moon, the whole face is illuminated. Between Quarter and Full Moon, the Moon is said to be gibbous.

Phobos The larger and closer-orbiting of the two moons of Mars. It is an irregularly shaped lump of rock some 27 km long and 20 km across (17 miles by 12 miles). It orbits at a distance of about 9400 km (5800 miles). Phobos is probably a captured asteroid.

Phoebe A tiny outer moon of Saturn, which has a retrograde orbit nearly 13,000,000 km (8,000,000 miles) away. It may be as small as 50 km (30 miles) across.

Phoenix A southern constellation near Achernar.

Phosphorus An ancient name for the planet Venus when it appears as a morning star.

photography See astrophotography.

photometer An instrument that measures the intensity of light or other radiation. Photoelectric photometers make use of photocells. Photomultiplier tubes are often used to amplify the faint light received. More sensitive measurements of brightness are now made with various forms of electronic image tubes.

photon The quantum, or fundamental unit of electromagnetic radiation, including light.

photon rocket A hypothetical rocket engine which uses a beam of light (i.e. photons) for propulsion. It features prominently in space fiction, but seems unlikely ever to be practical.

photosphere The visible surface of the Sun, which has a temperature of about 5500°C.

Piazzi, Giuseppe (1746–1826) Italian astronomer who discovered the first asteroid, Ceres, on the first day of the 19th Century.

Pictor The Painter; a southern constellation close to Carina.

Pioneer A series of US space probes, beginning in 1958. The most successful have been Pioneer 10 (Jupiter, 1973), Pioneer 11 (Jupiter, 1974; Saturn, 1979), and Pioneer Venus (Venus, 1978). Pioneer 10 became the first probe to journey into interstellar space when it crossed the orbit of Pluto in June 1983.

Pioneer plaque A 'message' to alien beings carried on Pioneer 10 and 11 in the form of an etched plaque, which tells who sent it and where we are located in the Galaxy.

Pisces The Fishes; a faint zodiacal constellation. Because of precession, it now contains the First point of Aries, one of the spots where the

Left: The changing appearance of the Moon as it goes through its phases every month. This natural cycle is used as the basis of lunar calendars.

Right: A Viking orbiter took this close-up picture of Phobos, the largest moon of Mars.

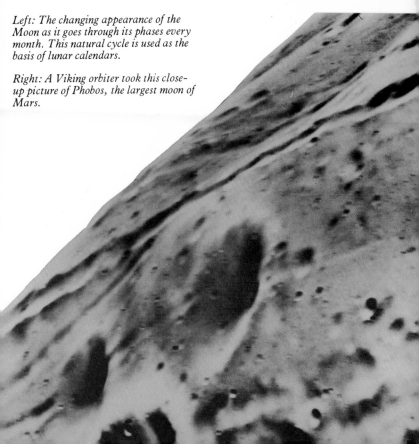

ecliptic crosses the celestial equator.

Piscis Austrinus The Southern Fish; a southern constellation dominated by its main star, the brilliant Fomalhaut.

plages See **flocculi**.

planet The major type of body in the solar system, which orbits the Sun. The word planet means 'wanderer', because the ancients thought that the planets were wandering stars. The Earth is one of nine planets and the third planet from the Sun. The others are, in order of increasing distance from the Sun: Mercury, Venus, (Earth), Mars, Jupiter, Saturn, Uranus, Neptune and Pluto. Pluto is the smallest planet, with a diameter of about 2500 km (1500 miles); Jupiter is the biggest, with a diameter of 143,000 km (88,700 miles). Venus is the planet that comes closest to Earth, approaching to within 42 million km (26 million miles). There is a marked difference in size and composition between the inner, or terrestrial planets (Mercury to Mars), which are small and rocky, and the outer planets, which are huge and gaseous (except for the tiny outermost planet, Pluto).

planetarium A building for demonstrating the motions of the heavenly bodies by means of a complicated projector. The term is also used for the projector itself, which projects images on to a domed ceiling to simulate the whole night sky. Originally, 'planetarium' was the name given to the device we now call an orrery.

planetary nebula This is a spherical nebula-like cloud that gives the appearance of a planetary disc when viewed through a telescope. The Ring nebula in Lyra is a famous example. It is a shell of gas 'puffed off' by a central very hot star as it passes from the red giant to the white dwarf stages in its evolution.

planetesimal Tiny bodies formed in the early stages of the solar system, according to a theory (early 1900s) by T.C. Chamberlain and F.R. Moulton, which eventually increased in size to become the planets.

planetoid
Another name for an asteroid.

Planet X The name given to a possible tenth planet that might exist beyond Pluto.

plasma A fourth state of matter (besides solid, liquid and gas) that is uncommon on Earth, but occurs in abundance in space. Matter in the searing hot interiors of stars is in the form of plasma. It is a state in which matter is split up entirely into ions and electrons. Atoms as such do not exist.

plate tectonics A field of study concerned with the movement of the plates that make up the Earth's crust which is responsible for continental drift.

Pleiades Also called the Seven Sisters; the most prominent open star cluster in the heavens, in the constellation Taurus. Although only about six or seven stars can be seen with the naked eye, the cluster actually contains several hundred stars.

Plesetsk
See **Northern Cosmodrome**.

Plough The most conspicuous part of the constellation Ursa Major which looks like an ancient plough, with ploughshare and handle.

Pluto The planet that orbits farthest from the Sun. At present and until 1999, however, it is actually travelling within the orbit of Neptune. It lies on average about 5900 million km (3670 million miles) from the Sun, and takes nearly 248 years to travel around the Sun once. It is the smallest planet, with a diameter of about 2500 km (1500 miles). It has one known moon, called Charon.

Opposite: The Helix nebula in the constellation Aquarius is a planetary nebula about 1.5 light-years across.

Right: Some of the bright stars of the Pleiades. They are young stars surrounded by nebulosity.

Pointers Two bright stars in the Plough (Alpha and Beta Ursae Majoris) which point towards the pole star, Polaris.

polar axis The axis of an equatorially mounted telescope that points to the celestial pole. The telescope is rotated around this axis to follow the path of the stars.

polar caps Patches of ice and snow visible at the north and south poles of Mars. They come and go with the Martian seasons.

Polaris (Alpha Ursae Minoris) The star that is at present the pole star, being located close to the north celestial pole. It is a 2nd magnitude binary star, one component of which is a Cepheid.

pole star A star near a celestial pole which appears to remain fixed in the sky. At present the pole star in the northern hemisphere is Polaris. There is no convenient pole star in the southern hemisphere.

Pollux (Beta Geminorum) A 1st magnitude star (K0) in the constellation Gemini, somewhat brighter than its 'twin', Castor. It lies 36 light-years away.

Poop See **Puppis**.

populations, star Two distinct types of stars, classified into Populations I and II by Walter Baade in 1944. Population I stars are relatively young and are found mainly in the spiral arms of galaxies. Population II stars are old and are found mainly in globular clusters and the centres of galaxies. They are not as rich in metals as Population I stars.

pore A tiny sunspot.

positional astronomy
  See **astrometry**.

Praesepe Also called the Beehive; an interesting open star cluster in Cancer, visible to the naked eye. It lies about 500 light-years away.

precession The gradual rotation of the Earth's axis in space, which causes the celestial pole to describe a circle in the heavens over a period of 25,800 years. This gives rise to the precession of the equinoxes – the gradual movement westwards of the points where the celestial equator intersects the ecliptic.

pressure suit A suit worn by pilots, and by astronauts as part of their spacesuit, to give them oxygen to breathe and to counter the lack of external pressure at high altitudes or in space.

prime focus The focal point of an objective lens or primary mirror. The prime-focus position in the big reflectors is often used for exposing photographic plates.

prime meridian The meridian through Greenwich, which is taken as the base line for the measurement of longitude on Earth.

primeval atom Also called fireball; the name often given to the concentration of matter, which exploded in the Big Bang to create the universe.

prism A wedge-shaped piece of glass used in some spectroscopes to split up incoming starlight into a spectrum for examination.

probe A spacecraft that escapes the Earth's gravity and travels into deep space to explore the planets, their moons, and interplanetary space. See **escape velocity**.

Left: A long-exposure photograph of the northern sky. The stars arc around the northern celestial pole, marked by Polaris.

Below: The Mariner 10 probe flew past Mercury twice, in 1973 and 1974, revealing a barren planet covered with craters.

Procyon (Alpha Canis Minoris) The eighth brightest star in the heavens, a mag 0.35 binary star that lies only about 11 light-years away.

Progress A robot ferry craft used by Russian space scientists to re-supply their space station, Salyut. Its design is similar to the manned Soyuz craft. It entered service in 1978.

prominence A great fountain of flaming gas from the Sun's chromosphere, which can rise to a height of hundreds of thousands of kilometres.

propellant The substance in a rocket that propels it. See **liquid propellant; solid propellant**.

proper motion The movement of a star across the celestial sphere, observable for only a few hundred of the closest stars. Bernard's star has the largest proper motion, of just over 10 seconds of arc per year. Proper motion does not represent the true motion of a star, merely the component of its motion at right-angles to our sight.

Prospero The only satellite launched by a purely British launch vehicle, Black Arrow, on 28 October 1971. It weighed 66 kg (145 lb).

proton A fundamental atomic particle present in the nucleus of all atoms. It has a positive electric charge. With a mass of about a million-million-million-million-millionth of a gram, it is 1836 times more massive than the electron.

proton-proton reaction One of the main nuclear fusion reactions responsible for providing the energy to keep a star shining. It is believed to be a three-stage reaction involving the fusion of protons (ordinary hydrogen nuclei) into the nuclei of heavy hydrogen and thence into helium nuclei. Overall, four protons are fused to form a nucleus of helium.

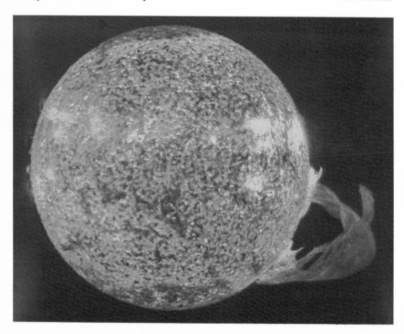

Proxima Centauri The star closest to the Sun, in the southern constellation Centaurus, at a distance of 4.28 light-years. It is a faint red dwarf star that belongs to the multiple star system Alpha Centauri.

Ptolemaic system The ancient Earth-centred view of the universe, as described by Ptolemy of Alexandria. The universe was bounded by a rotating celestial sphere holding the stars. In the centre was a fixed Earth. The planets revolved around the Earth in rather complicated circles known as deferents and epicycles. The system dominated astronomical thinking until the time of Copernicus in the 1500s.

Ptolemy of Alexandria Also called Claudius Ptolemaeus; the most influential astronomer of ancient times, who flourished in about AD 150. He summarized the astronomical knowledge of his day in an encyclopedic work, which we now know in its Arabic translation *Almagest* ('The Greatest').

pulsar An object that emits short rapid pulses of radiation at radio, X-ray or visible light wavelengths. It is thought to be a rapidly rotating neutron star, which gives off radiation in a rotating beam. We see a 'pulse' when the beam sweeps past our line of sight. The first pulsar was discovered at Cambridge in 1967 by Jocelyn Bell, working with Anthony Hewish. The most rapid-flashing pulsar is found in the Crab nebula. The Crab pulsar flashes on and off 30 times a second.

Opposite: Skylab astronauts photographed this enormous solar prominence in 1973.

Below: The Ptolemaic view of the universe in which the heavenly bodies circled around the Earth.

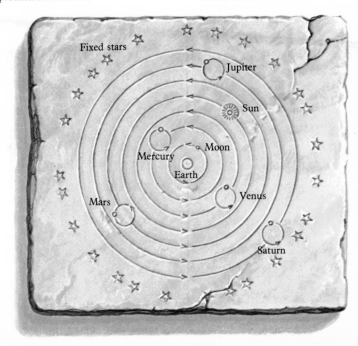

Puppis The Poop; a large southern constellation covering a rich region of the heavens and containing many fine star clusters.

Pyxis The Compass; a small and insignificant southern constellation.

Q

quadrant An instrument for determining the altitude of a star. It consists of a quarter circle, over which an arm carrying a sight can move.

Quadrantids A regular meteor shower, with the radiant in the constellation Boötes. It occurs in early January.

quadrature The position of a planet or the Moon when it is at right-angles to the Sun, when seen from the Earth. The Moon is at quadrature in its quarter phases.

quantum The smallest unit of something, particularly radiation. The photon is the quantum of light.

quark A hypothetical particle that is thought to be the true fundamental atomic particle. Ordinary atomic particles (protons, neutrons, etc) are thought to be made up of combinations of different kinds of quarks.

quasars Quasi-stellar radio objects, or QSOs; mysterious heavenly objects first detected by their powerful radio emissions and then identified visually as star-like bodies. When their spectra are investigated, they show enormous red shifts, indicating that quasars lie at very great distances. They seem to be thousands of times smaller than galaxies, but pour out the energy of hundreds of galaxies! It is thought that the enormous energy of a quasar could come from the region of a massive black hole.

quiet Sun A period when there is little sunspot and flare activity on the Sun.

R

radar astronomy Astronomy carried out by using radar – transmitting pulses of short radio waves (microwaves) and receiving their echo. It is used to investigate meteors and also the planets. Venus has been mapped by radar, both from the

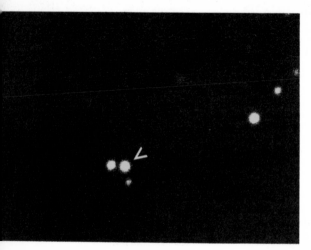

Right: This false-colour picture of Venus was built up from radar observations of the planet made by the Pioneer Venus orbiter.

Left: The quasar 3C-147, one of the star-like bodies that has the energy output of hundreds of galaxies.

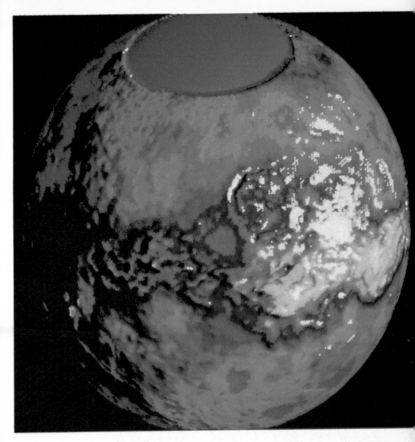

Earth and from orbit by space probes such as the Pioneer Venus orbiter.

radial motion Or radial velocity; the component of a star's movement towards or away from us. We cannot see it but we can detect and measure it in the shift in the star's spectral lines. See **blue shift; red shift**.

radiant The point in the heavens from which the meteors in a meteor shower appear to come.

radiation
See **electromagnetic spectrum**.

radiation belts
See **Van Allen belts**.

radio astronomy Astronomy that studies the radio waves that reach the Earth from outer space. They come through a 'radio window' in the atmosphere. Bell telephone engineer Karl Jansky founded radio astronomy when he discovered cosmic radio waves in 1931. Now a major branch of astronomy, radio astronomy has been responsible for outstanding discoveries in recent years such as quasars and pulsars. Radio astronomers work with huge aerial arrays called radio telescopes. See **radio telescope**.

Above: Some of the dishes of the Westerbork Synthesis Radio Telescope in the Netherlands. In all it has 14 dishes, which make an effective dish 3 km (nearly 2 miles) across.

Below: The RATAN radio telescope at Zelenchukskaya in the northern Caucasus. It measures some 600 metres (1970 ft) across, and is made up of 895 tilting and turning aluminium plates.

radio galaxy A galaxy which gives off exceptional amounts of energy at radio wavelengths, typically a million times greater than that of a normal galaxy like our own.

radioisotope thermoelectric generator (RTG) A kind of nuclear battery used on space probes which produces electricity from the heat given off by a radioisotope.

radio telescope The aerial, or aerial array used to gather and focus radio waves from the heavens. The most common type is dish-shaped. It has a huge parabolic reflector that gathers the waves and focuses them on to an aerial suspended above. The signals are then fed to a suitably tuned radio receiver. There are huge dish telescopes at Arecibo in Puerto Rico; Effelsberg in West Germany; and Jodrell Bank in England.

Ram See **Aries**.

Ranger A series of US probes launched to photograph the Moon's surface before crash-landing. The first success came with Ranger 7, in 1964.

RATAN 600 A powerful Russian radio telescope in the Caucasus.

Reaction control system (RCS) The system in a spacecraft that controls its attitude. It fires clusters of tiny rocket thrusters to cause rotation about the three axes.

reaction engine A rocket; an engine that works on the reaction principle, see **Newton's third law**.

recurrent nova A nova that recurs, such as T Coronae Borealis.

red giant A late stage in the life cycle of a star like the Sun. A star expands into a red giant when its nuclear fuel is nearly exhausted. Red giants are of low density, and are typically 100 times bigger than the Sun.

Red Planet A common name for Mars, which appears reddish-orange in the sky. Close to, its surface is also rusty red.

red shift The shift in the spectral lines of starlight towards the red end of the spectrum when the star is travelling away from us. This effective lengthening of wavelength is due to the Doppler effect. The amount of shift is a measure of the speed at which the star is receding. The light from the galaxies indicates that they are nearly all receding, and the farthest ones are travelling fastest. This provides evidence of an expanding universe.

Red Spot A large red oval in the atmosphere of Jupiter, which has persisted for centuries. Some 28,000 km (17,500 miles) long, it is thought to be a gigantic storm centre.

Redstone A US rocket developed as a medium-range ballistic missile in the 1950s and used as a launch vehicle for the Mercury suborbital flights of Shepard and Grissom in 1961.

re-entry The moment when a spacecraft returning from orbit hits the atmosphere. Re-entry craft use the atmosphere for braking, and air friction causes considerable heating. To survive, the craft must have an effective heat shield.

reflecting telescope See **reflector**.

reflection nebula A bright nebula that shines by reflecting the light from nearby stars. There is nebulosity of this type around the young, hot stars in the Pleiades.

reflector A telescope that uses mirrors to gather and focus incoming starlight. A typical reflector uses a parabolic primary mirror to gather the light, which is then brought to a focus in various ways. It may be brought directly to a prime focus further up the telescope tube or be deflected by a mirror and brought to a focus outside the tube. See **Cassegrain focus**; **coudé focus**; **Newtonian telescope**. The world's largest reflector is the Zelenchukskaya, which has a 6-m (236-inch) diameter mirror.

refracting telescope See **refractor**.

refractor A telescope that uses lenses to gather and focus starlight. It contains a large objective lens, which

A Redstone rocket, with its Mercury capsule payload. Redstone burned alcohol and liquid oxygen propellants.

Below: The lens arrangement in a refractor. The eyepiece is moved back and forth for focusing.

Objective

Eyepiece

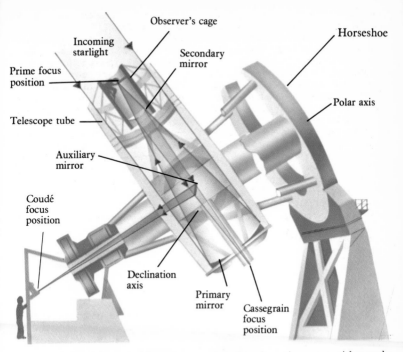

Incoming starlight

Prime focus position

Telescope tube

Observer's cage

Secondary mirror

Horseshoe

Polar axis

Auxiliary mirror

Coudé focus position

Declination axis

Primary mirror

Cassegrain focus position

The construction of a typical large reflector. A parabolic primary mirror gathers incoming starlight and reflects it back up the tube. It can then be brought into focus in various ways, either at the prime focus or elsewhere, using other mirrors.

forms an image, and a smaller eyepiece lens to magnify that image. The lenses are housed in a telescopic tube so that they can be moved relative to one another for focusing. The biggest refractor is the Yerkes, which has a 40-inch (102-cm) diameter objective.

regenerative cooling A method used to cool the exhaust nozzle of space rockets. It involves pumping cold propellant through the double wall of the nozzle.

regolith The surface 'soil' of the Moon and Mars, which on the Moon has been likened to having the consistency of ploughed farmland.

Regulus (Alpha Leonis) A brilliant 1st mag star (B7) in the constellation Leo. It lies about 85 light-years away.

relativity A concept developed by Albert Einstein that revolutionized ideas about energy, matter, space and time. The special theory of relativity was put forward in 1905. One of its consequences is that nothing can travel faster than the velocity of light. The theory also showed that energy and mass are equivalent, this being summed up in the famous equation $E = mc^2$. One consequence of the general theory of relativity (1916) is that gravitation causes light to bend, and this has been observed to happen.

remote manipulator system The space shuttle orbiter's 'crane', a 15-m (50-ft) long flexible arm that stows inside the payload bay. It was built by Canada.

remote sensing Taking pictures or images from a distance. Landsat is a very successful remote-sensing satellite, which scans the Earth's surface at various wavelengths. It produces informative false-colour images, which reveal a wealth of detail. Probes carry out remote sensing of distant planets.

resolving power The ability of a telescope to separate close stars or surface details in other bodies.

Reticulum The Net; a small southern constellation near Dorado.

retrobraking Slowing down a spacecraft by firing retrorockets.

retrograde motion Motion in the opposite direction from usual. Most moons revolve around their planets in an anticlockwise direction. Phoebe, the outer moon of Saturn, orbits clockwise – it has a retrograde orbit.

retrorocket A rocket fired in the direction in which a spacecraft is travelling so as to slow it down.

Rhea Saturn's second largest moon, a pitted ball of rock 1530 km (950 miles) across. It orbits at a distance of about 530,000 km (330,000 miles).

Ride, Sally (born 1951) The first US woman astronaut, who made her space début on the seventh shuttle flight, lifting off on 18 June 1983. She next returned to space on mission 41-G in October 1984, when she was accompanied by Kathryn Sullivan.

Rigel (Beta Orionis) A brilliant blue-white supergiant star (B8) of mag 0.11, the brightest star in the constellation Orion and the seventh brightest in the night sky. It lies over 800 light-years away.

right ascension One coordinate in the equator system of stellar location. It is the equivalent of terrestrial longitude. It is the distance along the celestial equator eastwards from the First point of Aries to the hour circle of the star. It is usually expressed in hours, minutes and seconds.

Rigil Kent See **Alpha Centauri**.

rille A trench-like feature on the Moon's surface. Rilles can run for hundreds of kilometres. Some connect chains of small craters.

rima An alternative name for rille.

ring mountain A typical lunar feature, being a large crater circled by high steep walls. Often there is a central mountain range (as in Copernicus).

Ring nebula (M57) A famous planetary nebula in the constellation Lyra, which looks like a smoke ring.

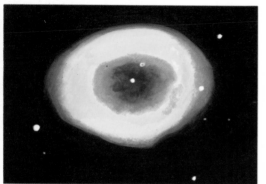

Left: The Ring nebula in Lyra. The ring of gas was puffed out by the central star about 6000 years ago.
Right: Dale Gardner hitches a ride on the remote manipulator arm during a satellite recovery mission in November 1984.

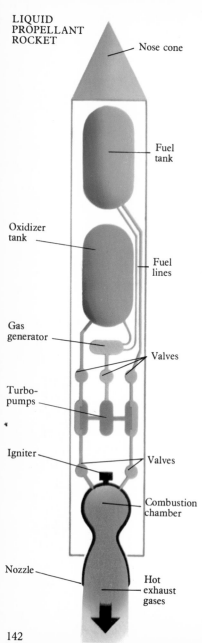

LIQUID
PROPELLANT
ROCKET

Nose cone

Fuel tank

Oxidizer tank

Fuel lines

Gas generator

Valves

Turbo-pumps

Igniter

Valves

Combustion chamber

Nozzle

Hot exhaust gases

Roche limit The minimum distance at which a moon can travel around its planet without disintegrating. It is 2.4 times the radius of the planet. This was discovered by the French astronomer E.A. Roche (1820–1883).

rocket An engine that is self-contained and is propelled by reaction. It carries both fuel, and oxygen to burn the fuel. It is thus independent of an external supply of air and can therefore work in space. The hot gases produced by burning the fuel are directed out of a nozzle and escape at high speed. Reaction to the gases escaping backwards propels the rocket forwards. Rockets may burn solid or liquid propellants (fuel and oxidizer).

rockoon A rocket that is carried high into the air by a balloon before it is fired.

Rømer, Ole Christensen (1644–1710) Danish astronomer who made one of the earliest accurate estimates of the velocity of light by timing the eclipses of Jupiter's moons by the planet.

Rosette nebula A beautiful spherical emission nebula in the constellation Monoceros, measuring about 50 light-years across.

Rosse, Third Earl of (1800–1867) William Parsons; British astronomer who established an observatory at Birr Castle in Parsonstown, Ireland. In 1854 he built there a 72-inch (183-cm) reflector, by far the biggest in the world at the time.

Royal Aircraft Establishment (RAE) Site of the British National Remote Sensing Centre, at Farnborough, Hants. It processes images obtained from Landsat and NOAA satellites.

The basic features of a liquid-propellant rocket. The propellants are pumped separately into the combustion chamber, where they mix and burn to form hot gases.

Royal Greenwich Observatory
(RGO) A world-famous observatory originally sited at Greenwich, near London. It was founded by King Charles II in 1675 for the purposes of improving navigation, and the meridian through Greenwich became the zero line for measurement of longitude. In 1948 the RGO was relocated at Herstmonceux Castle in Sussex, where there are much better viewing conditions. But most observations are now carried out at outstations, such as the new observatory at La Palma in the Canary Islands.

RR Lyrae stars Old short-period stars found in globular clusters, named after the prototype RR Lyrae in Lyra. They have a period of less than 1 day. They are also known as short-period Cepheids and cluster variables.

Rule See **Norma**.

Russell, Henry Norris (1877–1957) US astronomer best known as the co-originator of the Hertzsprung-Russell diagram.

Ryle, Martin (born 1918) British radio astronomer who pioneered many new techniques, including aperture synthesis for improving resolution.

Some of the most important launching vehicles of the Space Age. Of these, only Ariane and Soyuz are still operational.

Soyuz 49 m (160 ft)

111 m (365 ft)

Mercury-Atlas 30.8 m (94 ft)

Gemini-Titan 33.2 m (109 ft)

Vostok 38 m (125 ft)

Ariane 47.5 m (155 ft)

Saturn V

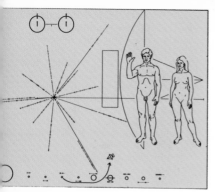

Ryumin, Valery (born 1939) Veteran Russian cosmonaut who is one of the most travelled men in space. Between February 1979 and October 1980 he spent a total of nearly 12 months in orbit in Salyut 6, in two missions of 175 and 184 days, which broke all previous space endurance records.

Top: Carl Sagan helped design the plaques that are carried by the Pioneer 10 and 11 probes to inform alien beings about Earth civilization.
Below: The Russian space station Salyut, docked with a Soyuz ferry craft. The whole complex measures about 20 metres (65 ft) long.

S

Sagan, Carl Edward (born 1934) American planetary astronomer particularly noted for his work on exobiology. He helped devise the 'messages' on the Pioneer and Voyager space probes (see **Pioneer plaque; Sounds of Earth**).

Sagitta The Arrow; a small constellation in the Milky Way between Cygnus and Aquila.

Sagittarius The Archer; a beautiful constellation of the zodiac. It abounds in rich star clouds, which lie in the direction of the centre of the Galaxy. It contains fine clusters and nebulae, including the Lagoon (M8) and the Trifid (M20).

Sagittarius A A source of powerful radiation in the centre of our Galaxy that could be the accretion disc of a black hole. It is thought to measure nearly 3 billion km (2 billion miles) across.

Sails See **Vela**.

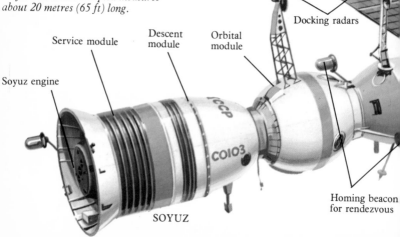

Docking radars

Service module

Descent module

Orbital module

Soyuz engine

Homing beacon for rendezvous

SOYUZ

Salyut A series of Russian space stations, first launched in 1971. The most successful have been Salyut 6, launched in 1977, and Salyut 7, launched in 1982 and still in orbit three years later. This latest design has multiple docking ports, allowing two or more craft to dock with it at any time. Cosmonauts are ferried up to the station in Soyuz spacecraft and are re-supplied periodically by unmanned Progress craft. In 1984, cosmonauts spent 237 days in Salyut 7.

San Marco platform A launch platform sited off the coast of Kenya just south of the equator, managed by Italian space scientists.

Saros A period of 18 years 11 days, after which the Sun and the Moon return to the same relative positions in space. So, after such an interval, eclipses of the Sun and Moon will occur in the same order and at the same times.

sarsat A search and rescue satellite; one that carries equipment to relay signals from emergency beacons on planes and ships. A number of NOAA and Cosmos satellites have this capability.

satellite Or moon; a small body that orbits around a planet. All the planets except Mercury and Venus have satellites. Earth has 1 (the Moon); Mars has 2; Jupiter at least 17; Saturn at least 22; Uranus at least 5; Neptune at least 2; and Pluto has at least 1.

satellite, artificial A man-made moon, launched into orbit by rocket. The Russians launched Sputnik 1, the first artificial satellite (usually just called satellite), on 4 October 1957. Today more than 100 satellites are launched every year, mainly by Russia, the US, and Europe through the European Space Agency. Most are communications satellites and weather satellites.

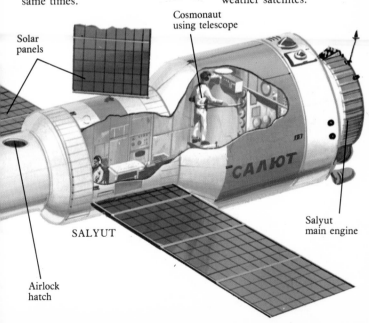

Cosmonaut using telescope

Solar panels

SALYUT

Airlock hatch

Salyut main engine

satellite galaxy A galaxy which is physically associated with a larger one. The two Magellanic Clouds are satellite galaxies of our own Galaxy.

satellite TV Television programmes that are relayed into the home from direct broadcasting satellites via local dish aerials.

Saturn The most beautiful planet in the solar system, because of the ring system around the equator (see **Saturn's rings**). The sixth planet out from the Sun, it orbits once every 29.5 years at a distance of 1430 million km (887 million miles). It is a gaseous giant of a planet, with a diameter of some 120,200 km (74,700 miles). It is the only planet with a relative density (0.7) less than water. It is made up mainly of hydrogen and helium. Saturn rotates rapidly on its axis, taking just over $10\frac{1}{2}$ hours to make one revolution.

Saturn nebula A planetary nebula in Aquarius that looks rather like Saturn, with 'rings' on either side.

Saturn rocket A series of US rockets used for manned space launchings. The most powerful, and the largest US rocket ever built, was the Saturn V, which was used to launch Apollo spacecraft to the Moon. Saturn V was 111-m (365-ft) long and had a take-off thrust of 3.4 million kg (7.5 million lb).

Saturn's rings A system of rings girdling the equator of Saturn. It measures about 272,000 km (169,000 miles) across. From the Earth we see three main rings – an inner faint C, or crêpe ring, a bright B ring, which is separated from the outer A ring by a dark gap (the Cassini division). There is a smaller gap near the edge of the A ring named the Encke division. From close up, the rings are

Saturn, seen from the surface of one of its moons. Box: Saturn's rings, seen from above. The dark band is Cassini's division.

seen to be made up of hundreds of separate ringlets.

Savitskaya, Svetlana (born 1948) Russian cosmonaut who became the second woman (after Tereshkova) to be launched into space, in August 1982. Exactly one year later she became the first woman to go spacewalking.

Scales See **Libra**.

Schiaparelli, Giovanni Virginio (1835–1910) Italian astronomer, whose reporting of 'canals' on Mars sparked off speculation about an intelligent Martian race.

Schmidt telescope A wide-field telescope invented by the Swedish-German astronomer Bernhard Schmidt (1879–1935). It uses a spherical mirror, and a correcting plate is placed in the telescope tube to correct distortions.

scintillation The twinkling of stars, caused by uneven refraction of their light through the atmosphere.

Scorpion See **Scorpius.**

Scorpius The Scorpion; a superb southern constellation that does look rather like its name. It is located almost entirely in the Milky Way, and its brightest star is the mag 1 red supergiant Antares. There are two fine globular clusters (M6 and M7) above the 'tail' that are visible to the naked eye.

Scout A small solid propellant US launch vehicle, used in numerous versions since 1960. It may have up to five stages.

Sculptor An inconspicuous southern constellation south of Cetus.

Scutum The Shield; a small constellation in the Milky Way just south of the celestial equator. It contains rich star fields and, near Beta Scuti, an open cluster (M11) called the Wild Duck.

Sea Goat See **Capricornus**.

sea, lunar See **mare**.

search and rescue satellite See **sarsat**.

Seasat A short-lived but highly successful US satellite launched in 1978, which gathered data about the shape of the Earth's surface by accurately measuring its altitude.

Sea Serpent See Hydrus.

seasons Regular annual variations in the climate, brought about because of the inclination of the Earth's axis ($23\frac{1}{2}°$) to the plane of its orbit around the Sun. As a result, some parts of the Earth are tilted more towards the Sun at some times than at others. The four seasons of winter, spring, summer and autumn are marked astronomically by the winter solstice, the vernal equinox, the summer solstice and the autumnal equinox.

seeing The state of the atmosphere for astronomical observations. There is good seeing on a clear calm night, bad seeing when the air is turbulent. E.M. Antoniadi devised a scale of seeing from I (good) to V (bad).

selenography Study of the Moon.

Serpens The Serpent; a large constellation separated by Ophiuchus into Caput (head) and Cauda (tail). In the east of the constellation, nearly on the celestial equator, there is a globular cluster (M5) that is just visible to the naked eye.

Serpent See **Serpens**.

Serpent-Bearer See **Ophiuchus**.

service module The part of the Apollo spacecraft that carried the main equipment, fuel cells, propellants and propulsion unit.

SETI Search for extraterrestrial intelligence.

Seven Sisters A common name for the Pleiades open star cluster.

Sextans The Sextant; a small, faint equatorial constellation south of Leo.

Seyfert galaxy First described by the US astronomer Carl Seyfert (1911–1960), a type of galaxy notable for its very bright nucleus, indicating an active core.

Shapley, Harlow (1885–1972) US astronomer noted for his investigation of globular clusters and the size and shape of the Galaxy.

SHAR India's main space launch site, on the island of Sriharikota, north of Madras.

Shaula (Lambda Scorpii) A bright (mag 1.62) blue-white star (B1) in the constellation Scorpius.

shepherd moons Tiny moons at the edge of Saturn's rings which apparently help keep the ring

Left: As part of their search for extraterrestrial intelligence (SETI) radio astronomers beamed this coded message to the star cluster M13 in 1974. It shows (left to right): the numbers 1 to 10 in binary code; numbers representing the 'atoms of life' hydrogen, carbon, nitrogen, oxygen and phosphorus; the structure of DNA; the size of human beings; a map of the solar system; and the Arecibo aerial that beamed the message.

Below: Two minutes after lift-off, the shuttle rids itself of the solid rocket boosters.

particles in place.

Shield See **Scutum**.

shooting star See **meteor**.

Shuang-cheng-tzu Or 'East Wind' Centre; China's main space launch site, located in the Gobi Desert in Inner Mongolia.

shuttle The US space shuttle transportation system. It is a re-usable system consisting of three main items of hardware – the orbiter, solid rocket boosters (SRBs) and external fuel tank. Of these only the external tank is not used again. The orbiter carries the crew and payload. It lifts off the launch pad attached to the external

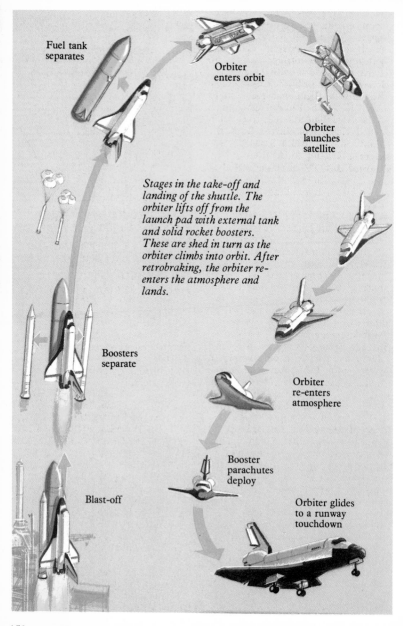

Fuel tank
separates

Orbiter
enters orbit

Orbiter
launches
satellite

Stages in the take-off and landing of the shuttle. The orbiter lifts off from the launch pad with external tank and solid rocket boosters. These are shed in turn as the orbiter climbs into orbit. After retrobraking, the orbiter re-enters the atmosphere and lands.

Boosters
separate

Orbiter
re-enters
atmosphere

Booster
parachutes
deploy

Blast-off

Orbiter glides
to a runway
touchdown

tank and the two SRBs. The SRBs soon separate and parachute back to Earth. Then just before orbit is reached, the external tank is jettisoned. The orbiter returns to Earth after the mission as a glider, landing on an ordinary runway. The first shuttle flight took place on 12 April 1981. The Russians are known to be developing a shuttle craft, often called Kosmolyot. In the West, the shuttle's main rival is Ariane.

sidereal Relating to the stars.

sidereal day The rotation period of the Earth with respect to the stars; it is 4 minutes shorter than a mean solar day.

sidereal time Time based on the rotation of the Earth in relation to the stars, ie on the sidereal day. Astronomers use sidereal time.

siderite An iron meteorite, which usually contains several per cent of nickel. When cut, polished and etched, it displays a characteristic Widmanstätten (triangular) pattern.

siderolite A meteorite containing both stony and metallic material.

Siding Spring Observatory An outstanding modern observatory in New South Wales, Australia, whose main instruments include the 3.9-m (154-inch) Anglo-Australian Telescope (AAT) and the 1.2-m (47-inch) UK Schmidt telescope.

signs of the zodiac See **zodiac**.

simulator An often complicated device that imitates, or simulates the actions of a machine. Shuttle astronauts train on the shuttle mission simulator, which is computer controlled and has 'live' instruments that react exactly as those of the real craft in real flight. Realistic scenes are projected on screens in front of the pilots.

singularity A hypothetical point in a black hole. According to theory, the matter from a heavy star's core is crushed into a singularity as it collapses under irresistable gravity.

Star trails arc around the southern celestial pole in this long-exposure photograph of the Siding Spring Observatory.

Sinope The outermost telescopic satellite of Jupiter, less than 50 km (31 miles) across, and circling more than 23,500,00 km (14,500,000 miles) out.

Sirius (Alpha Canis Majoris) Also called the Dog Star; the brightest star in the night sky, with a magnitude of -1.45, located in the constellation Canis Major. It is a white star (A1) of medium surface temperature (10,000°C) and lies relatively close to us, at a distance of 8.8 light-years. It is a binary star, whose other component is the so-called companion of Sirius, or Sirius B. This is a dim star, which was the first to be recognized as a white dwarf.

Top: Skylab in orbit in 1973. The gold-coloured patch covers an area damaged during launch, when a solar panel was ripped off.

Right: A Skylab astronaut tests a manned manoeuvring unit.

Skylab An experimental US space station launched in May 1973 and occupied over a period of 10 months by three teams of three astronauts, for respectively 28, 59 and 84 days. Astronauts were ferried up to Skylab by Apollo spacecraft. With this attached, the space station measured over 36 m (120 ft) long and weighed 90 tonnes. Skylab was built around a surplus Saturn rocket casing and was launched into orbit by a Saturn V rocket. It came out of orbit and broke up over the Indian Ocean in July 1979.

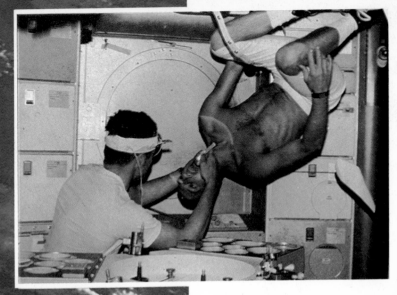

On Skylab Dr Joe Kerwin checks out Pete Conrad's mouth during one of the crew's regular medical examinations.

Skynet A series of British military communications satellites operating in geostationary orbit. Skynet IV is scheduled for launch from the shuttle in 1986 with Britain's first astronaut on board, Nigel Wood.

Slipher, Vesto Melvin (1875–1969) US astronomer who discovered the rotation of the galaxies.

Small Magellanic Cloud (SMC) Also called Nubecula Minor; the smaller of the two Magellanic Clouds, which are satellites of our Galaxy. Located in the far south in the constellation Tucana, the SMC lies about 200,000 light-years away and measures some 20,000 light-years across.

solar Relating to the Sun.

solar activity Disturbances occurring on the Sun. They include sunspots, flares, flocculi and prominences.

solar cell A semiconductor device that makes electricity directly from sunlight. Solar cells are made from doped wafers of silicon, rather like microchips. Practically all satellites are powered by solar cells.

solar cycle An alternative name for the **sunspot cycle.**

solar eclipse An eclipse that occurs when the Moon passes in front of the Sun and casts a shadow on the Earth. When the Sun's disc is completely obscured, the eclipse is total. When the disc is only partly obscured, the eclipse is partial. On average there are two or three solar eclipses a year. See also **annular eclipse; Baily's beads; corona.**

solar electric propulsion An ion rocket which generates the electricity to accelerate the ions from banks of solar cells.

153

Solar Max Popular name for the Solar Maximum Mission satellite, which was launched in February 1980 but failed after 10 months in orbit. In April 1984, however, shuttle astronauts recovered and repaired Solar Max in orbit and relaunched it, working perfectly.

solar panel A flat panel containing sets of solar cells. Solar panels extend like wings from many spacecraft.

solar power satellite A huge satellite, circling in geostationary orbit, which would capture solar energy, turn it into microwaves and beam them down to Earth. There they would be converted into electricity and fed into the ordinary grid system. The satellite would have to be many kilometres across in order to collect sufficient sunlight.

solar sail A novel concept to provide an interplanetary spacecraft with extra propulsion in its cruise phase. The idea is that the craft will unfurl a huge sail, hundreds of kilometres across, which would catch the feeble solar wind coming from the Sun and so be propelled.

solar system The Sun and its family of nine planets, at least 50 moons, and thousands of asteroids, comets and other lumps of rock and ice. The whole family is bound together by the Sun's enormous gravitational attraction, which is powerful enough to keep the outermost planet Pluto in its orbit at a distance of 7200 million km (4500 million miles). The solar system is about 5000 million years old and was probably formed from a nebula of gas and dust.

Left: Flying a manned manoeuvring unit, George Nelson closes on Solar Max during the recovery mission in April 1984.

Above: The McMath solar telescope at Kitt Peak Observatory. The solar image is reflected down the inclined shaft.

solar telescope A telescope for forming images of the Sun. An image is captured by a system of mirrors (heliostats, or coelostats) and then projected on to an observation table. The heliostat system is mounted on a tall tower, and the projection room is located at or below ground level. The world's biggest solar telescope is the McMath, at Kitt Peak Observatory, Arizona.

solar wind A stream of charged atomic particles that 'blows' steadily from the Sun, becoming stronger at times of increased solar activity. It consists mainly of protons and electrons.

solid propellant A rocket propellant made up of solid materials, in which fuel and oxidizer are mixed together. Gunpowder was the earliest solid propellant and is still used in fireworks rockets. Modern solid propellants often use a powdered aluminium as fuel and ammonium perchlorate as oxidizer, in a synthetic rubber binder.

solid rocket booster (SRB) The 45-m (149-ft) long solid propellant rocket used to help launch the shuttle. A pair of SRBs is strapped to the external tank at lift-off, together developing nearly 2.7 million kg (5.8 million lb) of thrust.

Solovyov, Vladimir (born 1946) One of the three Russian cosmonauts (with Atkov and Kizim) who established a new space endurance record in 1984 by staying in orbit in Salyut 7 for 237 days.

solstices Literally 'Sun standing still'; the times of the year when the Sun reaches its maximum northerly or southerly position among the stars. In other words, when the Sun has reached its greatest declination $23\frac{1}{2}°N$ (summer solstice) or $23\frac{1}{2}°S$ (winter solstice).

Left: The Soyuz 19 spacecraft, which took part in the Apollo–Soyuz Test Project in July 1975.

Right: The pressurized laboratory module of Spacelab.

Below: ASTP astronauts toast each other with tubes of space food, wishfully labelled 'vodka'.

Sounds of Earth A disc record that is being carried into interstellar space on the space probes Voyager 1 and 2 for the edification of any aliens who may find it in the future. It carries in code recordings of a variety of natural and man-made sounds heard on Earth, and greetings in 60 different languages. There are instructions about how to play it on the cover.

Southern Cross See **Crux**.
Southern Crown
  See **Corona Australis**.
Southern Fish See **Piscis Austrinus**.
southern lights
  The aurora australis; see **aurora**.
Southern Triangle See **Triangulum Australe**.
Soyuz The main Russian manned spacecraft since 1967, now used to

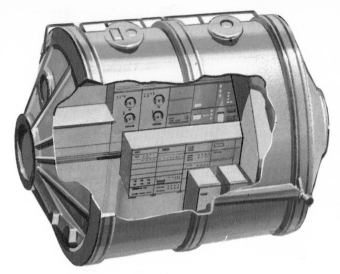

ferry cosmonauts to Salyut space stations. It is a modular craft some 7.9 m (26 ft) long. The crew return to Earth inside a spherical descent module, which is braked first by retrorockets and then by parachutes to a landing on the ground.

space The part of the universe that lies beyond the Earth's atmosphere, or more generally the volume that comprises the universe, in which all the heavenly bodies move.

space adaptation syndrome The correct term for space sickness.

space colony A human habitat established off the Earth. It is envisaged that next century permanent bases will be set up on the Moon and large cities could be built in space, from materials mined on the Moon. These cities would probably be of torus design (wheel-shaped) or cylindrical (see **O'Neill cylinders**).

spacecraft Any craft that travels in space, such as a satellite, which orbits the Earth, or a space probe, which escapes the Earth's gravity and travels into deep space.

space food is much more appetizing than it once was. Early foods were gooey pastes packed in toothpaste-type tubes. On the shuttle now some 70 different kinds of foods are available. Much is dehydrated, being mixed with water before serving. Other foods are canned or frozen or in natural form.

Spacelab A re-usable space laboratory that operates from inside the payload bay of the space shuttle. Designed and built by the European Space Agency (ESA), Spacelab is made up of pressurized laboratory modules and unpressurized instrument-carrying pallets. The first Spacelab flight took place in November 1983.

space medicine The branch of medicine concerned with the behaviour of the human body under the stresses of space flight, which include high G-forces at lift-off followed by weightlessness in orbit. Worrying problems about long-duration flights include bone-tissue loss and wasting muscles.

space probe See **probe**.

space shuttle See **shuttle**.

space sickness Properly called space adaptation syndrome; nausea experienced by many astronauts during the first few days of a space flight.

space station A large spacecraft in permanent orbit in which astronauts can work for long periods. The US Skylab was a successful experimental space station. The smaller Russian Salyut space stations have also proved successful for limited periods. A much larger US modular station is now being designed for permanent occupation in the 1990s.

spacesuit A multilayer garment worn by astronauts when they go spacewalking, that is, embark on extravehicular activity (EVA). The major layers are water-cooled 'long johns' next to the skin; a pressure suit to provide oxygen and apply pressure to the body; and external protection against extremes of cold and heat and harmful radiation. Early spacesuits were supplied with life support (oxygen, power etc) through a tube (umbilical) from the spacecraft. The shuttle spacesuit has a built-in life-support backpack.

space telescope Properly called the Edwin Hubble space telescope; an 11-tonne astronomical observatory scheduled for launch in 1986. Observing at visible wavelengths, it has a mirror 2.4 m (94 inches) across.

space transportation system (STS) A name often given to the space shuttle system.

spacewalk The popular term for what is correctly called extravehicular activity (EVA), or activity outside the confines of a spacecraft. Cosmonaut Alexei Leonov made the first spacewalk in March 1965. Cosmonaut Svetlana Savitskaya became the first woman to make a spacewalk in August 1984.

speckle interferometry A recent computerized technique developed to improve the resolving power of

Above: Skylab astronauts took part in many experiments in space medicine. Here Owen Garriott is seen in a lower body negative pressure device, having his blood circulation monitored.

optical telescopes. It involves the manipulation of tiny images called speckles. The first picture of the surface of another star (Betelgeuse) was built up by speckle interferometry.

spectral classes The groups into which stars are classified according to their spectra. See **Draper classification**.

spectral lines The lines observed in the spectrum of starlight. Usually they are dark absorption lines. The lines are like stellar 'fingerprints'.

Early next century space stations like this, made up of modules, could be in orbit.

They help to identify elements present in the star, since every element produces a characteristic set of lines. The spectral lines also provide a convenient method of classifying the stars.

spectrograph See **spectroscope**.

spectroheliograph An instrument developed by US astronomer George Hale, used to photograph the Sun in the light of one particular wavelength.

spectroscope An instrument for analysing light by means of its spectrum. It uses a prism or diffraction grating to produce a spectrum, which is then observed with a telescope. A spectrograph is a spectroscope with a camera attached to provide a photographic record (spectrogram) of the spectrum.

spectroscopic binary A binary star whose components are too close together to be separated visually, but reveal themselves by the shifting of lines in their spectra.

spectrum The spread of colour obtained when light is passed through a spectroscope. See **absorption spectrum; continuous spectrum; emission spectrum**.

spectrum The whole range of electro-magnetic radiations that bodies such as stars give out. See also **electromagnetic spectrum.**

spherical aberration A defect of lenses and curved mirrors, which causes blurring of the image. It happens because light rays from different parts of the curved surface are brought to a focus at slightly different points.

Spica (Alpha Virginis) A brilliant blue-white star (B1) in the constellation Virgo. At mag 0.96, it is the 15th brightest star in the sky.

spicules Short-lived spiky jets of gas seen erupting from the chromosphere of the Sun.

spiral galaxy One of the main types of galaxy, of which our own Galaxy is an example. Spiral (S) galaxies have a bulging nucleus from which radiate spiral arms. They are graded a, b and c according to the development of the spirals. An Sc spiral has widely separated arms. Our own Galaxy is an Sb spiral.

splashdown The landing at sea of a spacecraft. Until the coming of the space shuttle, all US space capsules splashed down at sea.

sporadics Ordinary meteors, which can be seen most nights of the year at the rate of up to about 10 an hour.

Spot A French Earth-resources satellite, launched by Ariane, which is able to take high-resolution images of the Earth's surface.

spring equinox
See **vernal equinox.**

spring tide A high tide that occurs at times of New and Full Moon, when the gravitational attractions of the Sun and the Moon are additive.

sputnik The Russian term for a satellite. Sputnik 1 was the world's first artificial satellite, when it went into orbit on 4 October 1957. An aluminium sphere some 58 cm (23 inches) in diameter, it weighed 84 kg (184 lb). It remained in orbit for just 92 days. The 500-kg (1120-lb) Sputnik 2 was launched a month later, and carried the world's first space traveller, a dog called Laika.

spy satellite A military reconnaisance satellite, designed to photograph military bases and operations of potentially unfriendly powers. See **Big Bird.**

Square of Pegasus The distinctive square made by four stars in the constellation Pegasus.

SRB See **solid rocket booster.**

stage One rocket unit of a multistage, or step rocket.

Left: This fine spiral galaxy, M83, lies about 27 light-years away. Our own Galaxy would probably look much like this from such a distance.

Below: Just after splashdown, the Apollo 11 astronauts wait to be picked up. Note their protective clothing, designed to prevent contamination of Earth.

star A gaseous celestial body, like the Sun, which shines by its own light. Light is one of the many ways in which a star gives off the energy produced by nuclear reactions in its interior. A star also gives off electromagnetic radiation of other wavelengths, from gamma-rays to radio waves.

Star City The English name for the Russian city of Zvezdnyy Gorodok, near Moscow, the location of the Gagarin centre where Russian cosmonauts undergo their training.

star cloud A region of the Milky Way where there is an exceptionally high density of stars. Star clouds abound in the constellation Sagittarius, towards the centre of the Galaxy.

star streaming An effect discovered in 1904 by J.C. Kapteyn that the stars have two preferential motions. This results from the rotation of the Galaxy.

star wars A popular term for wars that may be fought in space. They involve the use of weapons such as killer satellites, or antisatellite satellites (ASATs). These killer satellites may simply knock out their targets' electronics by swamping them with radiation; or else they may blast their targets with high-energy lasers or particle beams, or destroy them by means of shrapnel.

THE BRIGHTEST STARS			
Star		*Magnitude*	
		App	Abs
Sirius	(CMa)	−1.45	1.4
Canopus	(Car)	−0.73	−4.7
Rigil Kent	(Cen)	−0.2	4.3
Arcturus	(Boo)	−0.06	−0.2
Vega	(Lyr)	0.04	0.5
Capella	(Aur)	0.08	−0.6
Rigel	(Ori)	0.11	−7.0
Procyon	(Cmi)	0.35	2.7
Achernar	(Eri)	0.48	−2.2
Hadar	(Cen)	0.6	−5.0
Altair	(Aql)	0.77	2.3
Betelgeuse	(Ori)	0.8	−6.0
Aldebaran	(Tau)	0.85	−0.7
Acrux	(Cru)	0.9	−3.5
Spica	(Vir)	0.96	−3.4
Antares	(Sco)	1.0	−4.7
Pollux	(Gem)	1.15	1.0
Fomalhaut	(PsA)	1.16	1.9
Deneb	(Cyg)	1.25	−7.3
Mimosa	(Cru)	1.26	−4.7

THE NEAREST STARS		
Star		*Distance (l-y)*
Proxima Centauri	(Cen)	4.28
Alpha Centauri A,B	(Cen)	4.3
Barnard's star	(Oph)	5.9
Wolf 359	(Leo)	7.6
Lalande 21185	(UMa)	8.1
Sirius A,B	(CMa)	8.8
UV Ceti A,B	(Cet)	8.9
Ross 154	(Sgr)	9.5
Ross 248	(And)	10.3
Epsilon Eridani	(Eri)	10.8

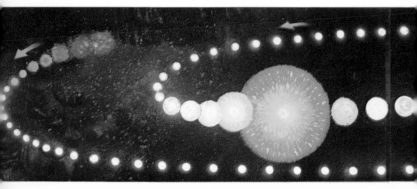

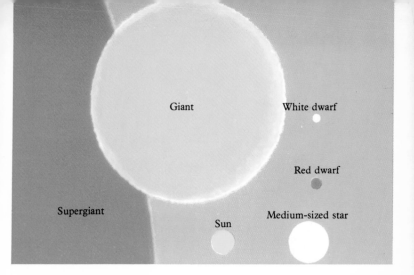

Giant

White dwarf

Red dwarf

Supergiant

Sun

Medium-sized star

stationary orbit
See **geostationary orbit**.

steady-state theory A theory about the origin and evolution of the universe, put forward by Hoyle, Gold and Bondi in 1948, but mostly out of favour now. It suggests that the universe has always been in the state it is today and will always remain so. To maintain the status quo, matter is continually being created. Compare **Big-Bang theory**.

stellar Relating to the stars.

stellar evolution varies from star to star, depending on its mass. Stars form from clouds of gas and dust, contracting under gravity and heating up until their nuclear furnaces 'fire'. A star like the Sun shines typically for about 10,000 million years, then expands into a red giant, probably goes through variable-star and planetary-nebula phases, and ends up as a white dwarf. More massive stars have a much briefer life, blowing themselves apart as a supernova, and then collapsing into a superdense neutron star or even further into the singularity of a black hole.

Top: Relative sizes of typical stars.
Left: A star like the Sun shines steadily for most of its life, before expanding into a red giant and then shrinking into a white dwarf.

163

step rocket Also called multistage rocket; a rocket vehicle made up of a number of separate rocket units (stages) joined together end to end. A single rocket by itself can never have a high enough power-to-weight ratio to reach orbital speed. The bottom stage of the step rocket fires first to boost the vehicle on its way. It separates and falls away when it runs out of fuel. The second stage then fires and boosts the now lighter rocket to a higher speed still, before separating in turn; and so on. In this way a favourable power-to-weight ratio can be achieved.

stinger An ingenious device used by shuttle astronauts to help recover two 'dead' communications satellites from orbit in November 1984. The astronauts inserted it in the propulsion motor nozzle of each satellite and then jetted with the satellite back to the orbiter.

Stonehenge An ancient stone circle near Amesbury in Wiltshire, which was probably built for astronomical purposes, such as determining the seasons and predicting eclipses. It dates from about 1800 BC.

stratosphere A layer in the atmosphere which extends from about 8–16 km (5–10 miles) above the Earth to a height of about 50 km (30 miles). Its temperature is steady at about –55°C.

STS Space transportation system; usually meaning the shuttle system.

summer solstice The time of the year when the Sun reaches its most northerly point in the heavens, on about June 21. In the northern hemisphere it is the longest day, and the beginning of summer. In the southern hemisphere, it is the shortest day and the beginning of winter.

summer time The standard time in Britain during the summer months (usually from the end of March to the end of October), which is one hour ahead of Greenwich Mean Time.

summer triangle A feature of summer skies in the northern hemisphere, a triangle formed by the bright stars Deneb (in Cygnus), Vega (in Lyra) and Altair (in Aquila).

Sun Our star; the body around which circle the Earth, the planets, and all the other members of the solar system. It lies on average about 149,000,000 km (93,000,000 miles) from the Earth. A globe of mainly hydrogen gas, the Sun is some 1,392,000 km (865,000 miles) in diameter, and is 333,000 times more massive than the Earth. It is a typical yellow dwarf star (G2) with a surface temperature of about 5500°C. Inside, its temperature rises to over 15,000,000°C. The Sun turns once on its axis in just over 24 days (at the equator). It travels once around the centre of the Galaxy in about 250 million years. Dark sunspots often appear on the bright surface of the Sun, the photosphere. The Sun's atmosphere – the chromosphere and corona – can normally be seen only at times of a total solar eclipse.

sundial A simple kind of clock that indicates time by the shadow cast by an indicator, or gnomon. It measures local solar time.

Left: The ancient stone circle of Stonehenge, the best preserved megalithic monument in the world.

Below: Inside the Sun, energy is produced in the core and travels slowly to the surface.

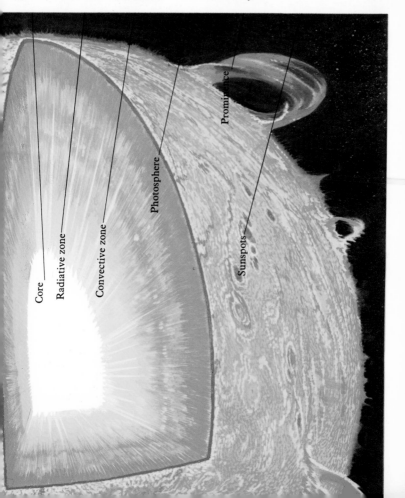

sunspot A dark blotch on the photosphere of the Sun, which may measure up to 200,000 km across. The largest ones may persist for months. It is a region that is some 1500°C cooler than the surroundings. Sunspots, which are usually associated with strong magnetic fields, come and go according to a regular solar, or sunspot cycle.

sunspot cycle Also called solar cycle; a periodic cycle during which the number of sunspots on the Sun varies from a minimum to a maximum and back again. The average cycle is about 11 years; there is evidence for a longer cycle of about 80 years. See **butterfly diagram**.

supergiant A huge star with a diameter typically hundreds of times that of the Sun. Supergiants, such as Betelgeuse and Antares, are very luminous, with high absolute magnitudes (to –7 and beyond).

superior conjunction
See **conjunction**.

superior planet A planet that lies beyond the Earth's orbit (ie Mars to Pluto).

supernova A star that suddenly flares up as it blasts itself apart. Its brightness may increase by as many as 20 magnitudes. Supernovae seem to occur in our Galaxy every few hundred years, the last being Kepler's star (1604). Many have been observed in other galaxies. As a result of a supernova explosion, a neutron star or a black hole may form.

supernova remnant An expanding gas cloud that is the visible remains of a supernova explosion. Among the best-known supernova remnants, which are often powerful radio and X-ray sources, are the Crab nebula and the Cygnus Loop.

Surveyor A series of US lunar probes, which soft-landed on the Moon between 1966 and 1968. They investigated the lunar soil and photographed the lunar landscape.

Swan See **Cygnus**.

Swordfish See **Dorado**.

Sword-Handle The double cluster in Perseus, h and Chi Persei, which is visible to the naked eye.

synchronous orbit
See **geostationary orbit**.

synchronous rotation
See **captured rotation**.

synchrotron radiation Electromagnetic radiation emitted when high energy particles such as electrons move through a strong magnetic field. This probably accounts for pulsar radiation, for example.

synodic period The time it takes a heavenly body to return to the same position in the sky, as viewed from the Earth. The synodic month, for example, is the time between two successive New Moons (29½ days).

syzygy A situation when three heavenly bodies are in a straight line. The Earth, Moon and Sun are in syzygy at New and Full Moon phases.

Above: These filaments of glowing gas form the supernova remnant we call the Veil nebula or the Cygnus loop. The supernova took place about 20–30,000 years ago.

Below: Two photographs of the galaxy NGC5253, the one on the right showing the presence in May 1972 of a particularly bright supernova.

T

T In astronautics, the time of lift-off of a launch vehicle. One hour before lift-off would be termed T-1; One hour after lift-off, T+1.

Table See **Mensa**.

Tanegashima The location of Japan's main rocket-launch site, an island at the southern tip of the country.

Tarantula nebula Also called the Looped nebula and 30 Doradus; a large bright nebula within the Large Magellanic Cloud. It is a region seething with violent activity from the radiation pressure and explosions of central stars.

Taurids A meteor shower, with radiant in Taurus, which occurs in late October and early November. It is thought to be associated with Encke's comet.

Taurus The Bull; a most interesting northern constellation of the zodiac. It contains two fine open clusters, the Pleiades and the Hyades, which is located near the fiery red giant Aldebaran. On the edge of the Milky Way in Taurus is the famous Crab nebula.

TDRS NASA's tracking and data relay satellite; a powerful communications satellite placed in geostationary orbit, which can relay signals back to a ground station from up to 25 different satellites at the same time. It has two 4.9-m (16-ft) dish antennae and measures 17.4 m (57 ft) across with its solar arrays deployed.

tektites Tiny glass-like objects found in certain parts of the world. It is thought that they may have originated in space, or were formed as a result of meteorite impact. See **australites; moldavites**.

telemetry Literally, measuring from a distance. The term is specifically used to describe the transmission and reception of instrument readings and other data between a ground station and spacecraft.

Telescope See **Telescopium**.

telescope An instrument for gathering and focusing light and other radiation coming from the heavenly bodies. Optical astronomers use lens and mirror telescopes to observe the heavens (see **refractor**, **reflector**).

Radio astronomers use **radio telescopes**. See also **solar telescope**; **space telescope**.

Telescopium The Telescope; an inconspicuous and uninteresting southern constellation.

Telstar 1 The first active communications satellite, launched by the US on 10 July 1962. It relayed the first live television programmes between the US and Europe during the 20-minute periods it was above the horizon for both continents.

Tereshkova, Valentina Vladimirovna (born 1937) Russian cosmonaut who became on 16 June 1963 the first woman to travel in space, in a Vostok capsule.

terminator The line that divides the dark and light hemispheres on a planet or moon.

terrestrial Relating to the Earth.

terrestrial planets Rocky planets like the Earth, namely Mercury, Venus and Mars.

tethered satellite A joint US/Italian programme to suspend a satellite into

Opposite: With the countdown clock at the Kennedy launch site showing T—9 minutes, the shuttle is ready for take-off.

Above: Telstar 1, the pioneering communications satellite.

Right: Valentina Tereshkova, the pioneering woman cosmonaut.

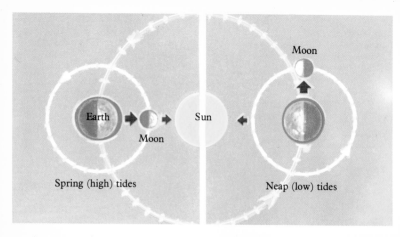

Spring (high) tides   Neap (low) tides

the upper reaches of the atmosphere from the shuttle on the end of a 100-km (60-mile) tether.

Tethys One of Saturn's moons, composed mainly of water ice with only a little rock. Measuring 1050 km (650 miles) across, it is crossed with deep cracks.

Thales of Miletus (*ca* 626–548 BC) Greek philosopher who is supposed to have predicted a solar eclipse, maybe that of 25 May 585 BC.

Tharsis Ridge A highland area straddling the equator of Mars, dominated by three massive volcanoes – Ascraeus Mons, Pavonis Mons and Arsia Mons. They rise to a height of about 20 km (12.5 miles).

thematic mapper (TM) An instrument flown on the US Landsat 4 and 5 Earth-resources satellites. It scans the Earth at seven different wavelengths. With a resolution of 30 m (96 ft), it can produce exceptionally detailed images.

Thor A first-stage US launch rocket, developed from the Thor intermediate range ballistic missile. It burned kerosene and liquid oxygen propellants. It was often used in combination with an Agena upper stage (Thor-Agena).

The individual tiles can be clearly seen in this picture of the orbiter Columbia.

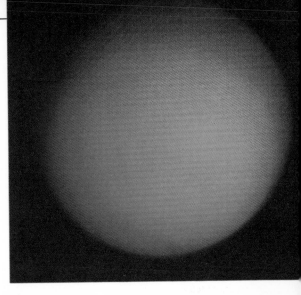

Right: Colourful Titan, largest of Saturn's moons. Uniquely for a moon, it has a thick atmosphere.

Left: The tides are caused by the attractions of the Moon and the Sun. These attractions may combine or oppose one another, causing high or low tides.

thrust The propulsive force developed by a jet or rocket engine.

tides The twice-daily rise and fall of the ocean, caused mainly by the gravitational attraction of the Moon, and to a lesser extent by that of the Sun. See also **neap tide**; **spring tide**.

tiles, shuttle The heat shield used to cover large areas of the shuttle orbiter, particularly the underside. Some 30,000 tiles are used, made of silica fibre. Each is individual and has to be fitted separately on to the airframe.

time Two main time periods are basic to the measurement of time – the day, the time it takes the Earth to rotate on its axis; and the year, the time it takes the Earth to make one orbit of the Sun (see **day**; **year**). The other divisions of time – seconds, minutes, hours, months – are arbitrary as far as civil timekeeping is concerned, although 'month' has a precise astronomical value (see **month**). See also **sidereal time**.

time dilation An effect predicted by relativity, that time passes more slowly the faster you travel. But this effect would only become noticeable at speeds approaching the velocity of light.

time zone One of 24 zones spaced throughout the world which have a standard time. Each zone covers roughly 15° of longitude. In each zone the time is exactly one hour behind the zone to the east, and one hour ahead of the zone to the west.

Tiros A series of US weather satellites, first launched in 1960. The latest of the series, the Tiros N type, is designated NOAA in orbit.

Titan By far the largest moon of Saturn, and the second largest (after Ganymede) in the whole solar system. Its diameter is 5140 km (3190 miles). It has a distinctive orange colour and is unique among moons in having an atmosphere, which is thicker than that of Earth. The atmosphere is made up mainly of nitrogen and methane.

Titania One of the five known moons of Uranus, about 1050 km (650 miles) across, which orbits 438,000 km (270,000 miles) out.

Titan rocket A powerful US launch vehicle which became operational in 1964. The Titan III configuration comprises a two-stage core vehicle burning nitrogen tetroxide and hydrazine, with twin solid rocket boosters. Coupled with a Centaur upper stage (Titan-Centaur) it is used for interplanetary missions (eg Viking, Voyager).

Titius-Bode Law Usually called **Bode's law.**

TM See **thematic mapper.**

Tombaugh, Clyde William (born 1906) US astronomer best known for his discovery of the ninth planet, Pluto, in 1930.

total eclipse An eclipse of the Sun when the Moon totally obscures the solar disc. Totality, the period of total darkness, can never last more than a little over 7 minutes.

Toucan See **Tucana.**

tracking Following the path of a rocket or spacecraft by means of radar, radio or photography. For tracking and communicating with space probes, huge dish aerials must be used to capture faint signals that may have travelled billions of kilometres. See also **Deep Space Network.**

tracking and data relay satellite See **TDRS.**

trajectory The path of a craft through the air or space.

trajectory correction manoeuvre (TCM) Altering the path of a space probe so as to bring it back on to its desired course. A number of TCMs are usually required for voyages into the outer solar system.

transfer orbit An orbit that takes a satellite from a low Earth orbit into a geostationary one.

transient lunar phenomena (TLP) Ruddy glows and outgassing occasionally observed in some craters on the Moon, including Aristarchus and Alphonsus, which suggest that the Moon is not entirely a dead world.

Transit A series of US navigation satellites, launched since 1968 specifically for the US Navy, but also widely used by merchant shipping.

transit The passage of one heavenly body across the face of another, for example, Mercury and Venus across the face of the Sun. This century transits of Mercury will occur in 1986, 1993 and 1999. The next transit of Venus will not occur until 2004.

transit, meridian The passage of a heavenly body across the meridian. The timing of a star's transit is done using a transit instrument (or meridian circle), a kind of telescope mounted on an east-west axis so that it always moves in the plane of the meridian.

Trapezium The multiple star Theta Orionis, embedded in the Orion nebula. The four brightest stars form the trapezium shape.

Triangulum The Triangle; a northern constellation which contains the fine galaxy M33, visible with binoculars.

Triangulum Australe The Southern Triangle; a far southern constellation close to the Milky Way.

Trifid nebula (M20) A brilliant and colourful nebula in the constellation Sagittarius, some 30 light-years across.

Triton The closer and larger of Neptune's two moons. About 4000 km (2500 miles) in diameter, it orbits 355,000 km (220,000 miles) out.

Trojans Two groups of asteroids linked gravitationally with Jupiter.

Tropics Circles on the celestial sphere and the Earth parallel with the equator. The Tropic of Cancer is $23\frac{1}{2}°$ north of the equator; the

The brilliant Trifid nebula is an emission nebula, producing light of its own. It is cleft with dark dust lanes, which are the birthplace of stars.

Tucana The Toucan; a large southern constellation which contains the Small Magellanic Cloud and the brilliant cluster 47 Tucanae.

47 Tucanae A fine globular cluster in the constellation Tucana which, at mag 5, is visible to the naked eye.

Tunguska event A catastrophic explosion that devastated some 50 square km (20 square miles) of forest at Tunguska, Siberia, on 30 June 1908. Eye-witnesses told of a huge fireball beforehand. Although once thought to be caused by a gigantic meteorite, it was more probably caused by the impact of the nucleus of a comet.

twinkling See **scintillation**.

Twins See **Gemini**.

Tycho One of the most spectacular lunar craters, not because of its size, which is modest (90 km, 55 miles), but because of the brilliant rays that emanate from it. They are

Top left: The far-sighted Konstantin Tsiolkovsky.

Below: The aftermath of the Tunguska event, which levelled trees like matchsticks.

Tropic of Capricorn, 23½° south. They mark the northernmost and southernmost declinations of the Sun every year.

troposphere The lowest layer of the Earth's atmosphere, this extends from 8–16 km (5–10 miles) according to latitude. It meets the second layer of the atmosphere, the stratosphere, at the tropopause.

Tsiolkovsky, Konstantin Edouardovich (1857–1935) Russian schoolteacher who worked out many of the basic principles of space flight, and is regarded as the 'father of astronautics'. He published his ideas first in his book *Exploring Space with Reactive Devices* (1903). He realized that only rockets would work in space; that powerful liquid-propellant rockets would be required; and that they would need to be joined together into a step rocket to achieve sufficient speed for space flight.

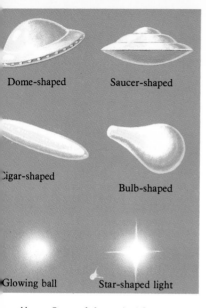

Dome-shaped   Saucer-shaped

Cigar-shaped

Bulb-shaped

Glowing ball   Star-shaped light

Above: Some of the typical forms UFOs take. The dome-shaped and saucer-shaped forms are the ones most commonly reported by people who have witnessed UFOs.

particularly noticeable at Full Moon. Tycho is located just north of Clavius in the southern hemisphere.

Tycho's star A supernova observed by Tycho Brahe in November 1572, which was bright enough to be seen in broad daylight. The expanding gas cloud from this supernova is hardly visible at optical wavelengths, but it shows up clearly at X-ray and radio wavelengths.

Tyuratam
See **Baikonur Cosmodrome**.

U

UFO Unidentified flying object. From the 1950s, mysterious flying objects have been reported throughout the world. Although popularly thought to be craft from other worlds, most UFOs turn out to be man-made objects (such as weather balloons and satellites) or natural phenomena (such as ball lightning). A few UFOs, though, still defy simple explanation.

U Geminorum A class of dwarf nova stars named after the prototype U Geminorum.

Uhuru A US scientific satellite launched from the San Marco platform off Kenya in 1970. Uhuru (Swahili for 'freedom') pioneered the exciting new field of X-ray astronomy.

ultraviolet astronomy Study of the heavens at ultraviolet (UV) wavelengths (shorter than visible light). Most UV wavelengths are blocked by the ozone layer in the atmosphere and so have to be studied from satellites. Among successful UV satellites have been Copernicus (1972) and the International Ultraviolet Explorer (IUE, 1978).

umbilical The term for the tube connecting an astronaut's spacesuit to the on-board life-support system of his spacecraft. It carries oxygen, cooling water, electricity, etc and incorporates a life-line.

175

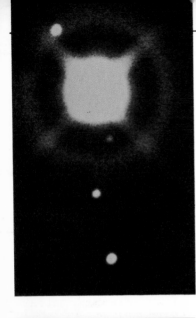

umbra The dark central region of a sunspot or the inner, totally dark shadow zone during an eclipse. Compare **penumbra**.

Umbriel The middle member of the five known moons of Uranus. Some 900 km (550 miles) across, it orbits at a distance of some 270,000 km (170,000 miles).

Unicorn See **Monoceros**.

Unidentified flying object See **UFO**.

Unisat A series of British direct broadcasting satellites (DBS), scheduled for launching from 1986 on from Ariane or the shuttle. They are designed to operate from geostationary orbit.

Universal time (UT) Standard civil time; the same as Greenwich Mean Time.

universe Everything that exists – space, matter, energy, radiation and so on. The universe is believed to be between about 15,000 and 20,000 million years old, and to have been created in a cataclysmic Big Bang that set it expanding. See **Big Bang**; **Big Crunch**; **closed universe**; **expanding universe**; **oscillating universe**.

uplink The transmission of signals up to a spacecraft.

Uranus The first 'modern' planet, unknown to the ancients. Discovered by William Herschel in 1781, Uranus is the seventh planet out from the Sun, and lies some 2870 million km (1783 million miles) away. It orbits the Sun every 84 years. With a diameter of 49,000 km (30,500 miles), it is one of the giant gaseous planets, probably resembling Jupiter

Top left: Uranus and the four closest of its five known moons.
Left: In the telescope Uranus appears vaguely green. A system of faint rings encircles the equator.
Right: A cutaway of the infamous V–2 rocket, the direct ancestor of all modern space rockets.

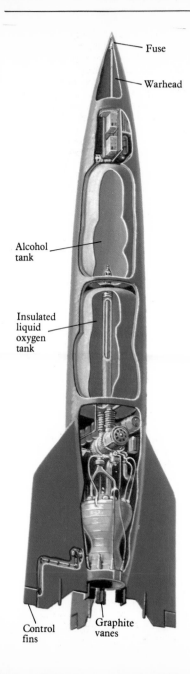

Fuse

Warhead

Alcohol
tank

Insulated
liquid
oxygen
tank

Control
fins

Graphite
vanes

and Saturn in composition. Unusually, the plane of Uranus's equator is nearly at right-angles (98°) to the plane of its orbit. Five moons and a system of faint rings girdle the planet.

Ursa Major The Great Bear; an unmistakable northern constellation whose main bright stars form the Plough. It is circumpolar from Britain. Mizar, the middle star in the 'handle' of the Plough, forms an optical double with Alchor. Merak and Dubhe (Alpha and Beta Ursae Majoris) act as pointers to the pole star, Polaris.

Ursa Minor The Little Bear; a far northern constellation whose main star, Polaris, is within one degree of the northern celestial pole and is presently the pole star.

Ursids A meteor shower that occurs just before Christmas, with radiant in Ursa Minor.

Utopia A plains region on Mars (Utopia Planitia) where the second Viking lander set down in 1976.

UV Abbreviation for ultraviolet radiation.

V

V-2 Hitler's second 'Vergeltungs-waffe', or revenge weapon of World War 2; a rocket bomb used to bombard London in 1944. Originally designated A-4, it was developed at Peenemünde on the Baltic Sea by a team headed by Wernher von Braun. After the War, captured V-2s provided the springboard for the missile and space programmes of the US and Russia. The V-2 was 14 m (45 ft) long; burned alcohol and liquid oxygen propellants; reached a height of about 96 km (60 miles); and had a range of some 300 km.

VAB See **Vehicle Assembly Building**.

Valles Marineris
See **Mariner Valley**.

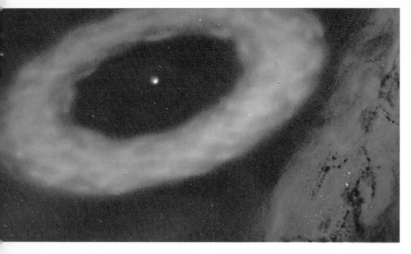

Above: The infrared satellite IRAS detected a ring of matter around the bright star Vega.

Below: An entry probe from Venera, designed to parachute into Venus's atmosphere.

Van Allen belts Two doughnut-shaped regions of intense radiation around the Earth. They occur where protons and electrons have become trapped in the Earth's magnetic field. The inner belt is about 3000 km high, the outer belt about 16,000 km high (2000 and 10,000 miles). They were discovered by the first US satellite, Explorer 1, and named after the person who conducted the experiment, James van Allen (born 1914).

Vandenberg Air Force Base A US base in California which is a major NASA space launch site. It is used particularly to launch satellites into a polar orbit. In 1985 the base also became operational as a second shuttle launch site, mainly for military launches. The launch site is also called the Western Test Range.

Vanguard A series of early US scientific satellites and launch vehicles, first launched in 1958.

variable star A star that varies in brightness. Intrinsic variables are stars that actually change in brightness because of processes going on inside them. The Cepheids are like this, as are RR Lyrae and Mira stars. Extrinsic variables are stars whose brightness varies because of an external influence, as in the case of eclipsing binaries.

Vega (Alpha Lyrae) The fifth brightest star (mag 0.04) in the heavens, in the constellation Lyra. It is a very hot blue-white star (A0) that lies about 26 light-years away. It forms one corner of the summer triangle. The satellite IRAS detected a ring of material around the star that could be the beginnings of another solar system.

Veha Two Russian probes launched in December 1984 to explore first Venus and then, in 1986, Halley's comet.

Vehicle Assembly Building (VAB) The building at the Kennedy Space Center where the various parts of the space shuttle are put together. The VAB, built originally to assemble the huge Saturn V Moon rockets, is 160 m high, 218 m long and 158 m wide (525 ft by 716 ft by 518 ft).

Veil nebula Part of the Cygnus Loop, a supernova remnant in the constellation Cygnus. It is a delicate tracery of brightly glowing gas.

Vela The Sails; a southern constellation set in the Milky Way, containing fine star clouds and clusters.

Venera A series of Russian space probes launched to investigate Venus. Venera 4 (1967) was the first to achieve success, parachuting an entry probe into the atmosphere. Venera 9 and 10 (1975) returned data and pictures after landing on the surface.

Venus A near twin planet of the Earth in size (diameter 12,140 km, 7545 miles), but very different in most other respects. It is shrouded in perpetual cloud. The atmosphere is mainly carbon dioxide, and the

Columbia leaving the VAB.

surface atmospheric pressure is 100 times that on Earth. The surface temperature, because of the atmosphere's 'greenhouse' effect, is more than 450°C. Venus has a slow retrograde rotation, turning on its axis every 244 days, while orbiting the Sun every 225 days. The surface of Venus, mapped by radar, is mainly flat rolling plains, but has two main highland areas, called Aphrodite Terra and Ishtar Terra. Venus comes closer to the Earth than any other planet, to within nearly 40 million km (25 million miles). This, coupled with its white cloudy atmosphere, make the planet shine brilliantly as the morning or evening star.

vernal equinox Or spring equinox; the time of the year when the Sun is exactly over the celestial equator, moving north, and day and night are of equal length all over the world. This happens on about 21 March and

179

marks the beginning of spring in the northern hemisphere; the beginning of autumn in the south.

Very Large Array (VLA) One of the world's most powerful radio telescopes, sited near Socorro in New Mexico. It works on the principle of interferometry and consists of a Y-shaped arrangement of 27 separate 25-m (82-ft) diameter dish aerials. Each arm of the Y is 21 km (13 miles) long.

Very Long Baseline Interferometry (VLBI) A technique used in radio astronomy to synthesize a huge radio telescope by processing data obtained by small radio telescopes dotted around the world. Theoretically a radio telescope the diameter of the Earth could be created.

Vesta The third largest of the asteroids, discovered by H.W.M. Olbers in 1807. Some 540 km (340 miles) across, it can sometimes be seen with the naked eye.

VfR Verein für Raumschiffahrt; a pioneering society for the study of space travel founded in Germany in 1927. An ex-member of the VfR was Wernher von Braun, who directed the development of the V-2.

Viking A US Mars probe, composed of two modules, an orbiter and a lander. Two Viking probes arrived at Mars in 1976, and their landers set down in July (Viking 1) and September (Viking 2). They took close-up photographs of the surface at Chryse Planitia (1) and Utopia Planitia (2); reported on the Martian weather; and tested the soil, in vain, for signs of life.

Virgin See **Virgo**.

Virgo The Virgin; A large zodiacal constellation whose main star is the brilliant blue-white Spica. It is a region rich in galaxies, including the Virgo cluster.

Virgo cluster One of the largest clusters of galaxies known, containing at least 2500 members. It lies about 60 million light-years away.

visible light The electromagnetic radiation to which our eyes are sensitive. White light is made up of many different wavelengths, or colours (violet, indigo, blue, green, yellow, orange and red), which are revealed when the light is split up into a spectrum.

visual binary A binary star, whose components can be separated in a telescope. Compare **spectroscopic binary**.

VLBI See **Very Long Baseline Interferometry**.

Volans The Flying Fish; an unremarkable southern constellation south of Carina.

von Braun, Wernher (1912–1977) German-born US rocket pioneer, who led the team that developed the notorious V-2 rocket and later headed the US space programme.

Porthole, Antennae, Re-entry capsule, Ejection seat, Oxygen, nitrogen storage bottles, Final rocket stage

180

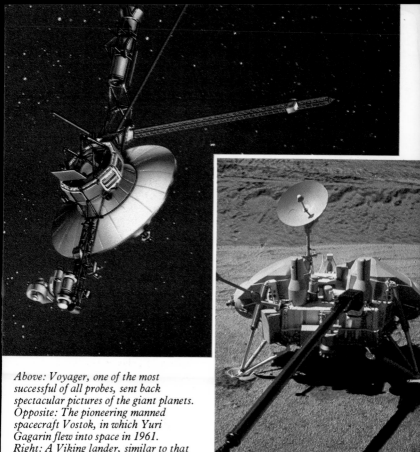

Above: Voyager, one of the most successful of all probes, sent back spectacular pictures of the giant planets.
Opposite: The pioneering manned spacecraft Vostok, in which Yuri Gagarin flew into space in 1961.
Right: A Viking lander, similar to that which landed on Mars in 1976.

Voshkod Russian manned spacecraft. Voshkod 1 flew in space in October 1964, carrying three cosmonauts. Voshkod 2 (March 1965) carried a two-man crew, including Alexei Leonov, who made the world's first spacewalk from it.

Vostok Russian manned spacecraft, the first of which carried Yuri Gagarin on the world's first manned space flight in April 1961. Five men and one woman (Valentina Tereshkova) made Vostok flights, the last in 1963.

Voyager US space probe launched to explore the outer solar system. Two probes were launched in 1977, both of which visited Jupiter and Saturn between 1979 and 1981, using the gravity-assist method to increase their speed. Voyager 2 then set course for encounters with Uranus (1986) and Neptune (1989). Both craft carry a record disc called 'Sounds of Earth'.

Vulcan The name given to a hypothetical planet once thought to orbit between Mercury and the Sun.

Vulpecula The Fox; an obscure northern constellation near Cygnus.

W

The beautiful Whirlpool galaxy

Wallops Flight Facility A NASA centre concerned with the assembly, launching and tracking of scientific satellites and sounding rockets. It is located on the Atlantic coast of Virginia.

wandering star The name the ancients gave to planets, which appeared to be stars that wandered across the heavens.

Water-Bearer See **Aquarius**.

Water immersion facility See **neutral buoyancy chamber**.

Water Snake See **Hydra**.

weather satellite One that monitors the weather by photographing and scanning the Earth's surface and atmosphere. Some satellites (such as NOAA) fly in polar orbits; others (such as Meteosat) fly in geostationary orbits.

weightlessness Also called zero-G; the sensation astronauts experience in orbit when they are in a state of free fall and appear to have no weight.

Western Test Range See **Vandenberg Air Force Base**.

Whale See **Cetus**.

Whipple, Frank Lawrence (born 1906) US astronomer noted particularly for his theory, now widely accepted, that comets are like 'dirty snowballs', made up of ice and dust.

Whirlpool galaxy (M51) A classic well-developed spiral galaxy (Sc) about 20 million light-years away in Canes Venatici, which we see face on. It is linked with a smaller companion galaxy and is exceptionally bright for its size.

White, Edward (1930–1967) The first US astronaut to make a spacewalk, during the Gemini 4 mission in June 1965. He stayed outside his craft for more than 20 minutes. Tragically, he was killed while training in an Apollo spacecraft, when it suddenly caught fire.

white dwarf A tiny white-hot star near the end of its life-cycle. Typically the size of the Earth but containing the mass of the Sun, it is incredibly dense. A teaspoonful would weigh over a tonne. The companion of Sirius was the first white dwarf to be so recognized.

white hole A hypothetical body which could be the place where the energy feeding into a black hole in this universe emerges in another universe.

White Sands Test Facility A NASA centre in New Mexico, used as an alternative shuttle landing site. It is also the ground station for receiving signals from the tracking and data relay satellites (TDRS).

white spot A feature that appears in the atmospheres of Jupiter and Saturn. It seems to be a storm centre.

William Herschel Telescope A British instrument installed at the Northern Hemisphere Observatory on La Palma, in the Canary Islands. Its light-gathering mirror is 4.2 m (165 inches) across, making it the third largest reflector in the world.

window The transparency of the atmosphere to incoming wavelengths from outer space. There are two main windows, the optical and the radio, which allow through visible light and radio waves. Most other radiation is absorbed partly or totally by the atmosphere.

window, launch
See **launch window**.

winter solstice The time of the year when the Sun reaches its most southerly declination, on about 23 December. It is the shortest day and the beginning of winter in the northern hemisphere; the longest day and the beginning of summer in the southern.

Wolf See **Lupus**.

Wolf-Rayet stars See **W stars**.

Below: Edward White floating in orbit during his spectacular spacewalk on 3 June 1965.

Woomera A British rocket launching site in South Australia, used for firing ELDO rockets and Black Arrow.

W stars or Wolf-Rayet stars; very bright hot stars of high absolute magnitude and surface temperatures of around 80,000°C.

W Virginis stars A class of Cepheid variable stars with a period of between 15 and 30 days. They are named after the prototype.

Below: The sleek X–15 rocket plane which flew so high that it had to manoeuvre like a spacecraft.
Opposite: John Young, pictured on the first flight of the shuttle Columbia in April 1981.

X

X-15 A US rocket plane that was able to reach a height of over 100 km (60 miles) and travel at Mach 6.

X-ray astronomy Studying the heavens at X-ray wavelengths. This must be done from space since X-rays are absorbed by the atmosphere. Uhuru pioneered X-ray satellite astronomy in 1970. Other major X-ray satellites have been Einstein (1979) and Exosat (1983).

Y

year Generally, the time it takes the Earth to make one revolution around the Sun. Astronomically several 'years' can be defined. The sidereal year is the period of revolution of the Sun with respect to the stars (=365.256 days). The anomalistic year is the time the Earth takes to circle the Sun from one perihelion to another (=365.259 days). The tropical (solar) year is the time the Earth takes to make one revolution in relation to the equinoxes (=365.242 days). The calendar year is 365.243 days.

Yerkes Observatory US Observatory located at Williams Bay, Wisconsin, near Chicago. It houses the world's largest refractor, which has an objective lens of 40 inches (102 cm) and is nearly 19 m (62 ft) long.

ylem A name for the original matter of the universe suggested by George Gamow.

Young, John (born 1930) Veteran US astronaut, who took part in two Gemini missions (III and X), two Apollo missions (10 and 16) and was commander on the first flights of the space shuttle and of Spacelab.

Z

Zeeman effect The splitting of the lines in a star's spectrum, caused by its magnetic field. Named after the Dutch physicist P. Zeeman, the effect provides a means of estimating the strength of the star's magnetism.

Zelenchukskaya telescope A Russian 6-m (236-inch) reflector located on Mount Semirodriki, near Zelenchukskaya in the Caucasus. It has been the world's largest optical

telescope since its completion in 1976. Unusually, it has an altazimuth mounting, not an equatorial one. It can detect objects as faint as the 25th magnitude.

zenith The point on the celestial sphere vertically above an observer.

zero-G Another term for weightlessness. It is misleading because gravity is still present in orbit. See **free fall**.

zodiac A circular band on the celestial sphere in which the Sun, Moon and planets appear to move during the year. It extends about 8° on either side of the ecliptic. The twelve constellations through which the zodiac passes are Aries, Taurus, Gemini, Cancer, Leo, Virgo, Libra, Scorpius, Sagittarius, Capricornus, Aquarius and Pisces. These constellations, or 'signs' of the zodiac, feature prominently in astrology.

zodiacal light A faint cone-shaped glow extending along the plane of the ecliptic. Observed usually in the tropics, it is caused by the scattering of light by interplanetary dust.

Zond A series of Russian space probes, the first of which was launched (unsuccessfully) to Venus in 1964. Most other Zonds were launched to the Moon, the later ones (1968–1970) returning to Earth afterwards.

Zvezdnyy Gorodok See **Star City**.

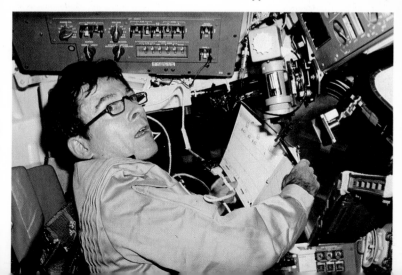

Constellations of the Southern Hemisphere

1 Capricornus, the Sea Goat
2 Aquarius, the Water-Bearer
3 Piscis Austrinus, the Southern Fish
4 Grus, the Crane
5 Tucana, the Toucan
6 Phoenix
7 Sculptor
8 Cetus, the Whale
9 Fornax, the Furnace
10 Eridanus
11 Lepus, the Hare
12 Orion
13 Monoceros, the Unicorn
14 Canis Major, the Great Dog
15 Columba, the Dove
16 Dorado, the Swordfish
17 Pictor, the Painter
18 Volans, the Flying Fish
19 Carina, the Keel
20 Puppis, the Poop
21 Pyxis, the Compass
22 Vela, the Sails
23 Leo, the Lion
24 Crater, the Cup
25 Hydra, the Water Snake
26 Corvus, the Crow
27, 28 Virgo, the Virgin
29 Libra, the Scales
30 Ophiuchus, the Serpent-Bearer
31 Serpens, the Serpent
32 Aquila, the Eagle
33 Sagittarius, the Archer
34 Corona Australis, the Southern Crown
35 Scorpius, the Scorpion
36 Lupus, the Wolf
37 Centaurus, the Centaur
38 Triangulum Australe, the Southern Triangle
39 Ara, the Altar
40 Indus, the Indian
41 Pavo, the Peacock
42 Apus, the Bird of Paradise
43 Octans, the Octant
44 Chameleon
45 Musca, the Fly
46 Crux, the Southern Cross
47 Hydrus, the Little Water Snake

2
31
30
33
34
35
36
29
28
39
40
41
38
37
27
42
43
46
26
45
44
47
25
24
18
16
17
19
22
21
23
15
20
11
14
12
13

187

Constellations of the Northern Hemisphere

1 Cetus, the Whale
2 Pisces, the Fishes
3 Aries, the Ram
4 Triangulum, the Triangle
5 Andromeda
6 Pegasus, the Flying Horse
7 Lacerta, the Lizard
8 Cygnus, the Swan
9 Equuleus, the Colt
10 Delphinus, the Dolphin
11 Aquila, the Eagle
12 Sagitta, the Arrow
13 Lyra, the Lyre
14 Draco, the Dragon
15 Hercules
16 Ophiuchus, the Serpent-Bearer
17 Serpens (Caput), the Serpent (Head)
18 Serpens (Cauda), the Serpent (Tail)
19 Corona Borealis, the Northern Crown
20 Boötes, the Herdsman
21 Coma Berenices, Berenice's Hair
22 Virgo, the Virgin
23 Canes Venatici, the Hunting Dogs
24 Ursa Minor, the Little Bear
25 Cepheus
26 Cassiopeia
27 Camelopardalis, the Giraffe
28 Ursa Major, the Great Bear
29 Leo Minor, the Little Lion
30 Leo, the Lion
31 Lynx
32 Cancer, the Crab
33 Hydra, the Water Snake
34 Canis Minor, the Little Dog
35 Gemini, the Twins
36 Orion
37 Taurus, the Bull
38 Auriga, the Charioteer
39 Perseus

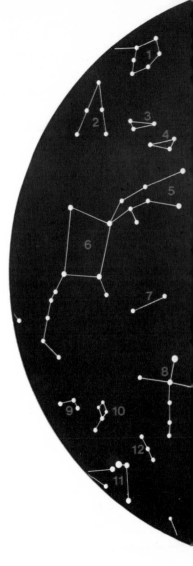

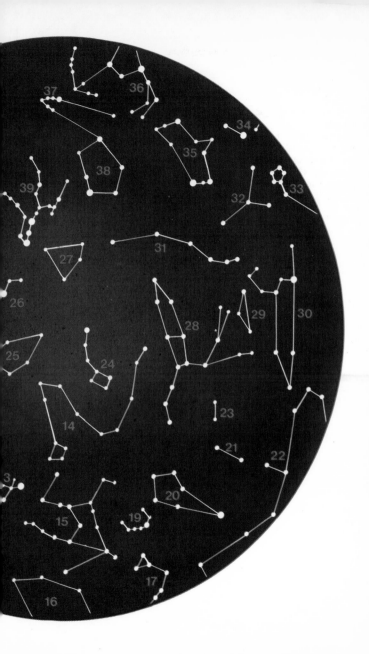

189

ACKNOWLEDGEMENTS
The author and publishers would like to thank the
following people and organizations for providing
the photographs included in this book. Picture
research by Spacecharts.

Allegheny Observatory
British Astronomical Association
British Interplanetary Society
EROS Data Center
European Space Agency
Grumman Corporation
Hale Observatories
Jet Propulsion Laboratory
Robin Kerrod
Kitt Peak Observatory
Adrian Lyons
Mullard Radio Astronomy Observatory
NASA
NASDA
Novosti
Parkes Radio Observatory
Royal Astronomical Society
Siding Spring Observatory
US Naval Observatory